PHYSICS

A Problem Solving Approach

Matthew McCluskey
Washington State University

Confocal Media
confocalmedia.com

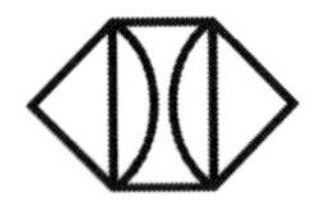

Confocal Media Edition, June 2015

Physics: A Problem Solving Approach

ISBN: 0692448918
ISBN-13: 978-0692448915
Print version 150618

Published by Confocal Media
confocalmedia.com

Send questions and comments for the author to:
mattmcc@alum.mit.edu

To my parents, John and Linda McCluskey

About the Author

Matthew McCluskey is a professor in the Department of Physics and Astronomy, Washington State University (WSU). He received a Physics B.Sc. from MIT in 1991 and Ph.D. from the University of California, Berkeley in 1997. He was a postdoctoral researcher at the Xerox Palo Alto Research Center (PARC) from 1997 to 1998. McCluskey joined WSU as an assistant professor in 1998. His research interests include semiconductors, high-pressure physics, and optics.

Professor McCluskey has taught Physics 101, WSU's introductory algebra-based physics course, numerous times. His text, *Physics: A Problem Solving Approach*, has been used by hundreds of students. He also co-authored a graduate text, *Dopants and Defects in Semiconductors*, and wrote a novel, *The Last Weapon*.

Preface

Physics is not easy. However, it *can* be simple.

This book is intended to help students learn basic physics principles, using a concise, problem-solving approach. It is a valuable resource for an introductory, algebra-based physics course. Hundreds of my students have used it successfully. It can supplement a traditional textbook or be used as a primary text.

To get maximum benefit, have a pencil and paper handy. If you own this book, feel free to write on the pages. Physics is about doing. The more you do, the more you learn.

The text is not intended to be comprehensive or highly detailed. Some topics, such as Newton's law of gravitation (the inverse square law), are omitted. The point is to emphasize what is most central to a first-semester physics course.

Overall, the chapters build on each other. However, the chapter on angular momentum (Chapter 10) can be skipped without impacting later chapters. The chapters on oscillations and waves (Chapters 12 and 13) can go after the fluids and thermal chapters (14 and 15).

In keeping with the taciturn style of this book, that's the end of the preface. Enjoy!

Units

Here is how we measure things:

Distance	meters (m)
Mass	kilograms (kg)
Time	seconds (s)
Angle	radians (rad) or degrees (°)

Some common conversions:

1 kilometer (km) = 1000 m = 10^3 m
1 centimeter (cm) = 0.01 m = 10^{-2} m
1 millimeter (mm) = 0.001 m = 10^{-3} m

1 gram (g) = 0.001 kg = 10^{-3} kg

1 minute (min) = 60 s

1 rotation = 2π rad = 360°

Examples:

1. Lauren ran 0.1 km, or 100 m, in 12 s. Her speed was 100 m / 12 s = 8.3 m/s.
2. Bob's mother in law has a mass of 1000 kg, or 10^6 g.
3. We can use dimensional analysis to convert units. Suppose a car goes 100 km/hr. We convert to m/s:

$$\frac{100\text{ km}}{\text{hr}}\frac{1\text{ hr}}{3600\text{ s}}\frac{1000\text{ m}}{\text{km}} = 27.8\text{ m/s}$$

Contents

1 Motion along one direction

Position and velocity

To describe the position of an object, we use an *axis*, which is like a number line:

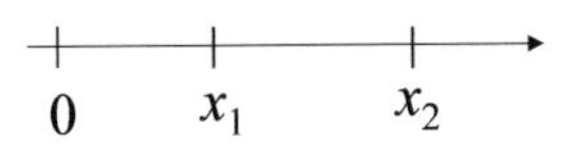

The 0 point is the *origin*. Suppose an object is at a position x_1, measured in meters. A while later, it is at position x_2. The *displacement* (Δx) is the difference between these two points:

$$\Delta x = x_2 - x_1$$

The time it takes to go from x_1 to x_2 is Δt. Assume the object's motion is steady, not slowing down or speeding up. The *velocity* (v) is measured in meters per second (m/s):

$$v = \Delta x / \Delta t$$

1. A car is at a position $x_1 = 10$ m and travels at a constant velocity. 3 s later, it is at $x_2 =$ 40 m. What is the displacement and velocity?

Displacement = Δx = 40 m – 10 m = **30 m**

Velocity = v = 30 m / 3 s = **10 m/s**

2. A plane is at a position $x_1 = 100$ km and travels at a constant velocity. 500 s later, it is at $x_2 = 50$ km. What is the displacement and velocity?

Displacement = Δx = 50 km – 100 km = **–50 km**

Velocity = v = –50 km / 500 s = **–0.1 km/s or –100 m/s**

Acceleration

Acceleration (a) is change in velocity (Δv) over change in time (Δt):

$$a = \Delta v / \Delta t$$

It has units of "meters per second, per second," or m/s^2.

If a is positive, then velocity is increasing. This is what happens to a car when you step on the accelerator pedal. If acceleration is negative, then you are slowing down, like when you slam on the brakes. If $a = 0$, then Δv is zero, which means that the velocity is constant.

Here is a plot of a car's velocity, where t is the time:

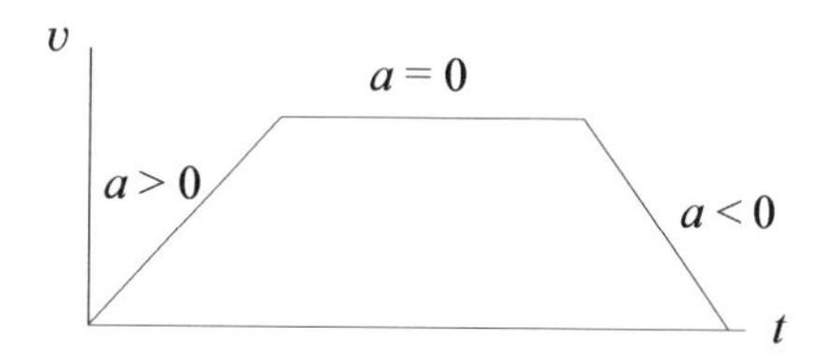

Initially, v is increasing, so acceleration is positive ($a > 0$). Later, v is constant, so $a = 0$. Finally, the car slows down, so acceleration is negative ($a < 0$).

1. A Mustang GT starts from rest. After 5 s, it has a velocity 20 m/s. Find the acceleration.

$\Delta v = 20$ m/s
$\Delta t = 5$ s
$a = 20$ m/s / 5 s = **4 m/s^2**

2. A plane travels with a velocity 40 m/s. After 10 s, it has a velocity 100 m/s. Find the acceleration.

$\Delta v = 100$ m/s – 40 m/s = 60 m/s
$\Delta t = 10$ s
$a = 60$ m/s / 10 s = **6 m/s^2**

Quiz 1.1

1. A runner is at a position $x_1 = 25$ m and has a constant velocity. After 9 s, she is at a position $x_2 = 70$ m. What is the displacement and velocity?

2. A ball is at a position $x_1 = -10$ cm and has a constant velocity. After 5 s, it is at a position $x_2 = -50$ cm. What is the displacement and velocity?

3. A missile starts from rest. After 10 s, it has a velocity 500 m/s. Find the acceleration.

4. A car travels with a velocity 20 m/s. The driver slams on the brakes, and 2 s later, the car has stopped. Find the acceleration.

Velocity as a function of time

We use equations to predict an object's position and velocity at some later time.

Suppose we have a stopwatch. We start at time $t = 0$. The position at $t = 0$ is the *initial position* (x_0). The velocity at $t = 0$ is the *initial velocity* (v_0).

At a time t, the velocity is given by

$$v = v_0 + at$$

1. The initial velocity of an object is 5 m/s and the acceleration is 3 m/s². What is the velocity at $t = 7$ s?

$v_0 = 5$ m/s
$a = 3$ m/s^2
$v = 5 + (3)(7) =$ **26 m/s**

2. An object starts from rest and accelerates at −7 m/s². Find the velocity after 11 s.

$v_0 = 0$
$a = -7$ m/s^2
$v = 0 + (-7)(11) =$ **−77 m/s**

Position as a function of time

The object's position is given by:

$$x = x_0 + v_0t + \tfrac{1}{2}at^2$$

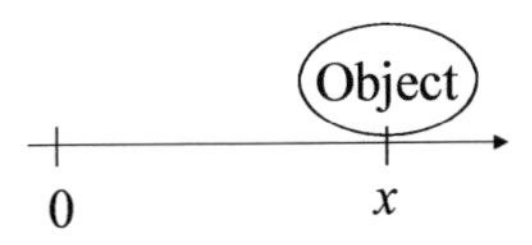

1. An object has an initial position 2 m, initial velocity 3 m/s, and acceleration 4 m/s^2. What is the position at $t = 5$ s?

$x_0 = 2$ m
$v_0 = 3$ m/s
$a = 4$ m/s^2

$x = 2 + (3)(5) + \tfrac{1}{2}(4)(5^2) =$ **67 m**

2. An object starts from rest and accelerates at 5 m/s^2. What is the displacement after 10 s?

Let's have the object start at the origin. We will often do this, because it makes life simple.

Since the object starts at the origin, $x_0 = 0$
The object starts from rest, so $v_0 = 0$
$a = 5$ m/s^2

$x = 0 + (0)(10) + \tfrac{1}{2}(5)(10^2) =$ **250 m**

Quiz 1.2

1. The initial velocity of an object is –5 m/s and the acceleration is 1 m/s². What is the velocity at $t = 5$ s?

2. An object starts from rest and accelerates at 10 m/s². What is the velocity after 3 s?

3. An object has an initial position –6 m, initial velocity –5 m/s, and acceleration 4 m/s². What is the position at $t = 5$ s?

4. An object starts from rest and accelerates at –0.1 m/s². What is the displacement after 10 s?

Two objects

Sometimes we want to track the position of two objects. We will write the equation of motion

$$x = x_0 + v_0t + \frac{1}{2}at^2$$

for each one individually.

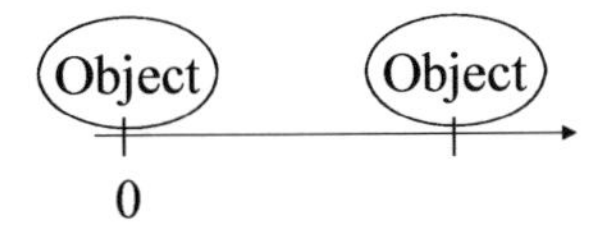

1. Kevin walks with a velocity 1 m/s. He is initially 300 m ahead of Rodney, who runs at a velocity 6 m/s. How long does it take for Rodney to catch up with Kevin?

Assume constant velocity ($a = 0$).

Kevin: $x = 300 + (1)t + \frac{1}{2}\,0t^2 = 300 + t$
Rodney: $x = 0 + (6)t + \frac{1}{2}\,0t^2 = 6t$

When Rodney catches up with Kevin, their x values are the same.

Set them equal: $300 + t = 6t$
$300 = 5t$
$\underline{\mathbf{60\ s}} = t$

2. A missile is launched at a jet, which is initially 250 m away. The missile accelerates from rest, with $a = 100\ \text{m/s}^2$. The jet has a constant velocity 200 m/s. How long does it take for the missile to impact the jet?

Missile: $x = 0 + 0t + \frac{1}{2}(100)t^2$
Jet: $x = 250 + (200)t + \frac{1}{2}\,0t^2$

Set them equal: $50\,t^2 = 250 + 200t$
$50\,t^2 - 200t - 250 = 0$

Quadratic equation: $t = \dfrac{-b \pm \sqrt{b^2 - 4ac}}{2a}$

$$t = \frac{200 \pm \sqrt{40{,}000 + 4(50)(250)}}{2(50)} = \frac{200 \pm 300}{100} = -1{,}5$$

Only the positive answer makes sense, so $t = \underline{\mathbf{5\ s}}$

Gravity

Gravity causes all objects to accelerate downward. This *gravitational acceleration* is denoted g. On earth, g = 9.8 m/s². This is very close to 10 m/s², so we will use that value:

$$g \approx 10 \text{ m/s}^2$$

Let's choose an axis that points up, toward the sky. Then, gravitational acceleration has a *negative* value, –10 m/s². That's because gravity points down, which is opposite to our axis.

Note: we always ignore air resistance.

1. An object is dropped from a height 20 m. How long does it take to hit the ground?

20 m — Object
0 — Ground

$x = x_0 + v_0t + ½at^2$

$x_0 = 20$ m
$v_0 = 0$ ("dropped" means the initial velocity is zero)
$a = –10$ m/s²
$x = 0$ when the object hits the ground. Find t.

$0 = 20 + 0t + ½ (–10)t^2$
$0 = 20 – 5t^2$
$5t^2 = 20$
$t^2 = 4$, so $t =$ **2 s**

2. A person throws a ball up, with an initial velocity 15 m/s. The ball goes up and comes back down. How long is the ball in the air before the person catches it?

Let $x_0 = 0$
$v_0 = 15$ m/s
$a = –10$ m/s²
$x = 0$ when the person catches the ball. Find t.

$0 = 0 + 15t + ½ (–10)t^2$
$0 = 15t – 5t^2$
$0 = 15 – 5t$
$5t = 15$, so $t =$ **3 s**

Quiz 1.3

1. A mouse runs away from a cat, at a velocity 2 m/s. The cat is initially 10 m away from the mouse. The cat runs toward the mouse at 7 m/s. How long does it take for the cat to catch the mouse?

2. A bank robber drives a car at a constant velocity 20 m/s. A super-fast police car is initially at rest, 800 m away. The police car accelerates at 20 m/s^2. How long will it take for the police to catch the robbers?

3. A ball is dropped from the roof of a building, 45 m above the ground. How long does it take for the ball to hit the ground?

4. A person shoots a bullet straight up. The initial velocity is 100 m/s. How long does it take for the bullet to hit the ground? (Ignore the height of the person.)

Chapter summary

Displacement (m)	$\Delta x = x_2 - x_1$
Velocity (m/s)	$v = \Delta x / \Delta t$
Acceleration (m/s²)	$a = \Delta v / \Delta t$
Velocity as a function of time	$v = v_0 + at$
Position as a function of time	$x = x_0 + v_0 t + \frac{1}{2}at^2$
Gravitational acceleration	$g = 9.8 \text{ m/s}^2 \approx 10 \text{ m/s}^2$

End-of-chapter questions

1. An ant moves with a constant velocity. It starts at $x_1 = 17$ mm. After 54 s, it is at $x_2 = -10$ mm. Find the displacement and velocity.

2. A soccer player runs with a velocity 8 m/s. The player slides to a stop in 2 s. Find the acceleration.

3. An object has an initial velocity –5 m/s and accelerates at 10 m/s². What is the velocity after 3 s?

4. An object starts from rest and accelerates at 10 m/s². What is the displacement after 10 s?

5. An object has an initial velocity 7 m/s and accelerates at 10 m/s². What is the displacement after 10 s?

6. In a race, Boris runs at a constant velocity 8 m/s. Suzie is initially 20 m behind Boris. Suzie runs at a constant velocity 10 m/s. How long will it take for Suzie to catch up with Boris?

7. A water balloon is dropped from the roof of a building, 125 m above the ground. How long does it take for the balloon to hit the ground?

8. An object is launched upward with an initial velocity 35 m/s. How long does it take for the object to land on the ground?

Challenging problems

1. An object accelerates from rest, with $a = 4$ m/s^2.

   a. How long does it take to travel 2 m?
   b. What is its velocity, after it has traveled 2 m?

2. Two cars approach each other. Initially, they are 375 m apart. The first car has an initial velocity 10 m/s and accelerates at 20 m/s^2. The second car has constant velocity –15 m/s.

   a. How long will it take for them to collide?
   b. What is the displacement of the first car, when the collision occurs?

3. A ball is dropped from a height of 5 m.

   a. How long does it take for the ball to hit the ground?
   b. What is the velocity of the ball, just before it hits the ground?

4. A ball is thrown upward, with an initial velocity 7 m/s. How long does it take the ball to reach its maximum height?

2 Motion in two dimensions

Cartesian coordinates

Before, we looked at motion along one dimension, x. Now, we are going to look at two dimensions, x and y. We plot an object's position (x, y) using Cartesian axes:

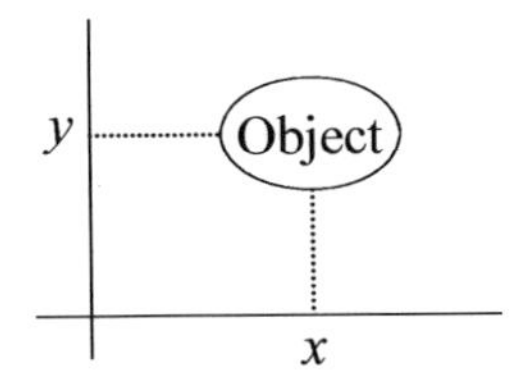

We need two numbers, x and y, to describe the position. Similarly, we need two numbers, v_x and v_y, to describe the velocity. Let's look at a constant velocity. An object's x value changes by Δx in a time Δt, while the y value changes by Δy.

The *x-component* of the velocity is

$$v_x = \Delta x / \Delta t$$

The *y-component* of the velocity is

$$v_y = \Delta y / \Delta t$$

1. A car is at a position (10 m, 15 m) and travels at a constant velocity. 3 s later, it is at (16 m, 3 m). What is the velocity?

$\Delta x = 16\text{ m} - 10\text{ m} = 6\text{ m}$
$\Delta y = 3\text{ m} - 15\text{ m} = -12\text{ m}$

$v_x = 6\text{ m} / 3\text{ s} = \underline{\mathbf{2\text{ m/s}}}$
$v_y = -12\text{ m} / 3\text{ s} = \underline{\mathbf{-4\text{ m/s}}}$

Velocity vector

The velocity is described by two numbers, v_x and v_y. We visualize the velocity by drawing a right triangle. The horizontal leg has a length v_x and the vertical leg has a length v_y:

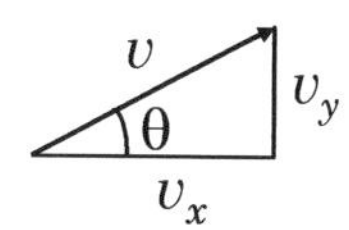

The hypotenuse is an arrow, called the *velocity vector*. The length of this arrow tells us the *speed*. The speed (v) is given by Pythagoras:

$$v = \sqrt{v_x^2 + v_y^2}$$

The direction that the vector points is the direction of motion. We can express direction by the *horizontal angle* θ. (It is also called "angle with respect to horizontal.") From trigonometry,

$$\cos\theta = v_x / v \qquad \sin\theta = v_y / v$$

1. An object has a velocity $v_x = 4$ m/s, $v_y = 3$ m/s. Find the speed.

$v = \sqrt{v_x^2 + v_y^2} = \sqrt{4^2 + 3^2} = \sqrt{25} = \underline{5 \text{ m/s}}$

2. An object has a speed $v = 8$ m/s. The vertical component of the velocity is $v_y = 4$ m/s. Find the horizontal angle.

$\sin\theta = 4 / 8 = 0.5$

$\theta = \sin^{-1}(0.5) = \underline{30°}$

3. An object has a speed $v = 50$ m/s and a horizontal angle $\theta = 60°$. Find the horizontal component of the velocity vector.

$\cos\theta = v_x / v$

$\cos(60°) = v_x / 50$
$0.5 = v_x / 50$

$\underline{25 \text{ m/s}} = v_x$

Quiz 2.1

1. An object is initially at the origin and has a constant velocity. After 15 s, it is at a position (30 m, –60 m). What is the velocity?

2. An object has a velocity $v_x = 12$ m/s, $v_y = 5$ m/s. Find the speed.

3. An object has a speed $v = 10$ m/s. The horizontal component of the velocity is $v_x = 5$ m/s. Find the horizontal angle θ.

4. An object has a speed $v = 88$ m/s and a horizontal angle $\theta = 30°$. Find the vertical component of the velocity vector (v_y).

Equations of motion

In the previous chapter, we used equations to predict the velocity and position of an object. For two dimensional motion, there are two sets of equations, one for x and one for y. We write a subscript x or y to tell them apart.

x:

$$v_x = v_{0x} + a_x t$$
$$x = x_0 + v_{0x} t + \tfrac{1}{2} a_x t^2$$

y:

$$v_y = v_{0y} + a_y t$$
$$y = y_0 + v_{0y} t + \tfrac{1}{2} a_y t^2$$

In these equations,

(x, y) is the position of the object at time t

(v_x, v_y) is the velocity at time t

(x_0, y_0) is the initial position ($t = 0$)

(v_{0x}, v_{0y}) is the initial velocity ($t = 0$)

a_x is the acceleration in the x direction

a_y is the acceleration in the y direction

1. An object is initially at the origin. It has a constant velocity $v_{0x} = 4$ m/s, $v_{0y} = 7$ m/s. Find the position after 8 s.

The object is initially at the origin: $x_0 = y_0 = 0$

Constant velocity means no acceleration: $a_x = a_y = 0$

$x = 0 + (4)(8) + \tfrac{1}{2}0(8^2)$ = **32 m**

$y = 0 + (7)(8) + \tfrac{1}{2}0(8^2)$ = **56 m**

Launching a projectile horizontally

Here is an example of projectile motion. An object is launched with an initial velocity that is horizontal. That means $v_{0y} = 0$.

Gravity causes the object to accelerate downward, with $a_y = -10$ m/s^2. Gravity does *not* cause acceleration in the horizontal direction, so $a_x = 0$.

We choose the axes so the object's initial position ($t = 0$) is at the origin: $x_0 = y_0 = 0$.

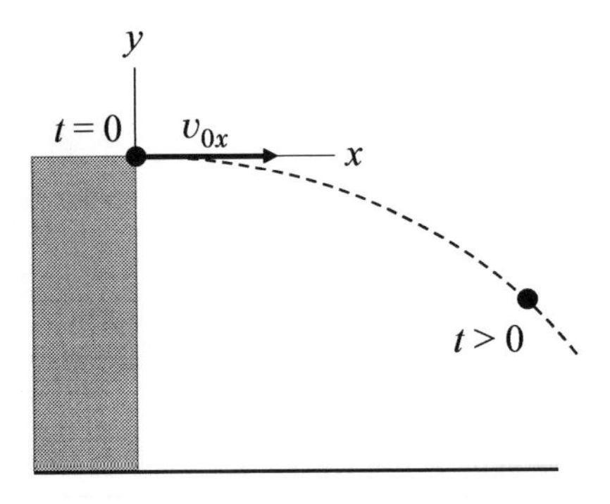

1. An object is launched horizontally from a 20 m tall building. The initial speed is 40 m/s. How long does it take for the object to hit the ground?

$y = y_0 + v_{0y}t + \frac{1}{2} a_y t^2$

When the object hits the ground, $y = -20$ m.

y is negative because of our choice of axes.

$-20 \text{ m} = 0 + 0t + \frac{1}{2}(-10)t^2$
$-20 = -5t^2$
$4 = t^2$, so $\underline{t = 2 \text{ s}}$

2. In the previous problem, what is the velocity just before the object hits the ground?

$v_x = v_{0x} + a_x t = 40 + (0)(2) = \underline{\mathbf{40\ m/s}}$

$v_y = v_{0y} + a_y t = 0 + (-10)(2) = \underline{\mathbf{-20\ m/s}}$

Quiz 2.2

1. An object is initially at (7 m, 8 m). It has a constant velocity $v_{0x} = 0$ m/s, $v_{0y} = -5$ m/s. Find the position after 11 s.

2. An object is launched horizontally from an 80 m tall building. The initial speed is 34 m/s. How long does it take for the object to hit the ground?

3. In the previous problem, what is the velocity just before the object hits the ground?

Launching a projectile at an angle

In this example, a gun points at a horizontal angle θ and shoots a projectile with initial speed v_0. The projectile will accelerate downward due to gravity. After some time, it will hit the flat ground. We choose the axes so the object's initial position is at the origin: $x_0 = y_0 = 0$.

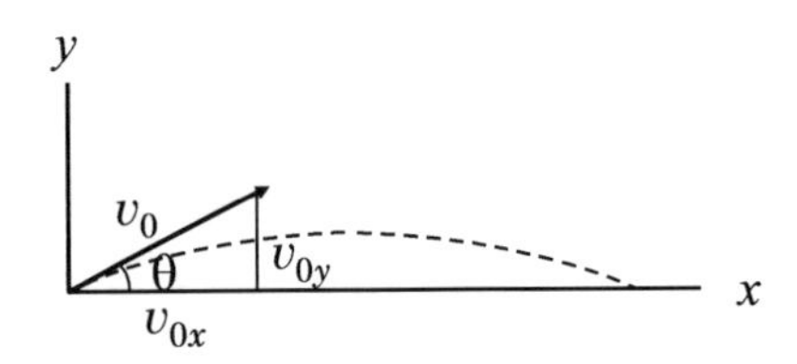

From trigonometry,

$$\cos\theta = v_{0x} / v_0 \qquad \sin\theta = v_{0y} / v_0$$

To solve for v_{0x} and v_{0y}, we multiply through by v_0:

$$v_0 \cos\theta = v_{0x} \qquad v_0 \sin\theta = v_{0y}$$

1. A cannon shoots a ball with initial speed v_0 = 40 m/s and horizontal angle θ = 30°. How long is the ball in the air before it hits the ground?

$y = y_0 + v_{0y}t + ½a_yt^2$

$v_{0y} = v_0 \sin\theta = (40\text{ m/s})(\sin 30°) = 20\text{ m/s}$

When the ball hits the ground, $y = 0$

$0 = 0 + (20)t + ½(-10)t^2$
$0 = 20t - 5t^2$
$0 = 20 - 5t$
$5t = 20$, so $\underline{t = 4\text{ s}}$

2. In the previous problem, what is the horizontal distance (x) that the ball travels?

$x = x_0 + v_{0x}t + ½a_xt^2$

$v_{0x} = v_0 \cos\theta = (40)(\cos 30°) = 34.64\text{ m/s}$

$x = 0 + (34.64)(4) + ½0t^2 = \underline{\textbf{139 m}}$

Hitting a target

Like before, a gun is aimed at a horizontal angle θ. It fires a projectile at a target that is some distance away. Because of gravity, the projectile will accelerate downward.

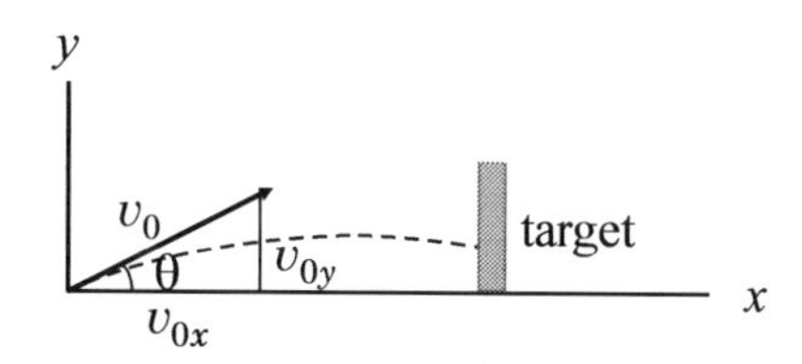

1. A cannon shoots a ball at a wall 200 m away. The ball has an initial speed 50 m/s and horizontal angle θ = 60°. How long does it take for the ball to hit the wall?

$x = x_0 + v_{0x}t + ½a_xt^2$

$v_{0x} = v_0 \cos θ = (50)(\cos 60°) = 25$ m/s
When the ball hits the wall, $x = 200$ m

$200 = 0 + (25)t + ½0t^2$
$200 = 25t$, so **$t = 8$ s**

2. In the previous problem, at what height (y) did the ball hit the wall?

$y = y_0 + v_{0y}t + ½a_yt^2$

$v_{0y} = v_0 \sin θ = (50)(\sin 60°) = 43.30$ m/s

$y = 0 + (43.30)(8) + ½(–10)(8^2) \quad = \quad$ **26.4 m**

Quiz 2.3

1. An archer shoots an arrow with an initial speed $v_0 = 14$ m/s and horizontal angle $\theta = 45°$. How long is the arrow in the air before it hits the ground? (Ignore the height of the archer).

2. In the previous problem, what is the horizontal distance that the arrow travels?

3. A sniper aims his rifle at a horizontal angle $\theta = 1°$ and fires a bullet with an initial speed 800 m/s. The target is a horizontal distance $x = 1600$ m away. How long does it take for the bullet to hit the target?

4. In the previous problem, what is the vertical position of the bullet (y) when it hits the target?

Chapter summary

Velocity vector

$$v_{0x} = v_0 \cos \theta$$
$$v_{0y} = v_0 \sin \theta$$

Equations of motion

x:

$$v_x = v_{0x} + a_x t$$
$$x = x_0 + v_{0x} t + \tfrac{1}{2} a_x t^2$$

y:

$$v_y = v_{0y} + a_y t$$
$$y = y_0 + v_{0y} t + \tfrac{1}{2} a_y t^2$$

End-of-chapter questions

1. A person shoots a rifle at a 45° angle. The initial speed of the bullet is 100 m/s. What is the initial velocity (v_{0x} and v_{0y})?

2. A college student throws a water balloon horizontally through an open window, with a speed 7 m/s. The window is 45 m above the ground. How long does it take for the balloon to hit the ground?

3. In the previous question, what is the velocity of the balloon just before it hits the ground?

4. A catapult launches a boulder at a 45° angle and initial speed 28.3 m/s. The ground is flat. How long does it take for the boulder to hit the ground?

5. In the previous question, how far does the boulder travel?

6. A person throws a dart at a horizontal angle 30° and initial speed 16 m/s. The dart hits a tree 7 m away. How long was the dart in the air?

7. In the previous question, the dart sticks in the tree some distance above the dart's initial height. What is that distance?

Challenging problems

1. A person is on the roof of a building 10 m above the ground. He throws a water balloon vertically upward, with initial velocity 5 m/s. How long does it take for the balloon to hit the ground?

2. An athlete throws a javelin at a 45° angle and initial speed 14.14 m/s. How long does it take for the javelin to reach its maximum height?

3. A cannon is on a hill that is 100 m above the flat ground. It shoots a cannonball at a horizontal angle 30° and initial speed 10 m/s.

   a. How long does it take for the ball to hit the ground?
   b. What horizontal distance (x) did the ball travel?

4. A biologist aims a rifle at a horizontal angle 60° and shoots a tranquilizer dart with an initial speed 10 m/s. A monkey is in a tree, a horizontal distance $x = 5$ m away. The dart hits the monkey. What was the height of the monkey (above the rifle)?

3 Forces

Force causes acceleration

Force is a vector that pushes or pulls on an object. It has units of kg m/s², also called Newtons (N).

A horizontal force acts along the x direction. If the force acts toward the right ($+x$ direction), it's positive. If it acts toward the left, it's negative. A vertical force acts along the y direction. Up is positive, and down is negative. The *magnitude* of a force is a positive number. For example, a horizontal force of –7 N has a magnitude of 7 N.

Forces cause objects to accelerate. Let F_x = the sum of forces along the x direction. Force and acceleration are related by

$$F_x = ma_x$$

Similarly, letting F_y = the sum of forces along the y direction,

$$F_y = ma_y$$

1. A rocket in outer space has a mass 1000 kg. Its engine produces a thrust of 20,000 N. What is the rocket's acceleration?

$F_x = ma_x$
20,000 N = (1000 kg) a_x
20 m/s² = a_x

2. Three forces act on a 1 kg mass. What is the acceleration?

1 N
1 kg
3 N
2 N

$F_x = ma_x$
3 N = (1 kg) a_x
3 m/s² = a_x

$F_y = ma_y$
1 N – 2 N = (1 kg) a_y
–1 N = (1 kg) a_y
–1 m/s² = a_y

Gravity

The force of gravity acts vertically downward. If the y axis points straight up, then the gravitational force is

$$F_y = -mg$$

where m is the object's mass and g = 10 m/s². Gravity does not act along the horizontal direction. The magnitude of the force is called the object's *weight*. Weight is a positive number.

1. Bob's mother in law has a mass of 1000 kg. What is the gravitational force on her? What is her weight?

$$\begin{aligned} F_y &= -mg \\ &= -(1000)(10) \\ &= \underline{\mathbf{-10{,}000\ N}} \end{aligned}$$

Weight = **10,000 N**

2. An object falls. What is the acceleration?

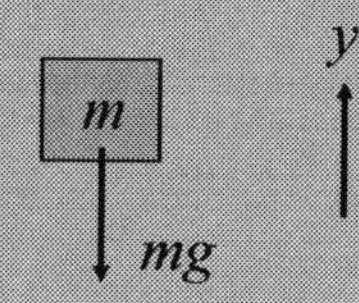

$$\begin{aligned} F_y &= ma_y \\ -mg &= ma_y \\ -g &= a_y \\ \underline{\mathbf{-10\ m/s^2}} &= a_y \end{aligned}$$

Notice how the mass canceled. Any object in free fall has the same acceleration, regardless of mass!

Quiz 3.1

1. A car's engine produces a horizontal force of 800 N. The car mass is 1000 kg. What is the acceleration?

2. Three forces act on an object of mass 7 kg. What is the acceleration?

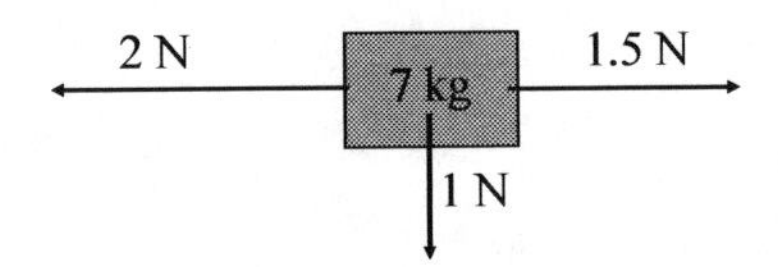

3. A block has a mass of 8 kg. What is the gravitational force on the block? What is the block's weight?

4. Galileo drops two masses, 1 kg and 8 kg, from the leaning tower of Pisa. What is the acceleration of each mass?

Forces between two objects

Suppose a person pulls a crate in the positive x direction, with a force of 30 N.

Look at the force on each object. The crate experiences a force $F_x = +30$ N. The person experiences an *equal but opposite* force $F_x = -30$ N.

1. A truck pulls a trailer. The truck exerts a force $F_x = 700$ N on the trailer. What force does the trailer exert on the truck?

   $F_x =$ **-700 N**

2. A person pushes an elephant with a force $F_x = 20$ N. What force does the elephant exert on the person?

   $F_x =$ **-20 N**

Tension

When a rope pulls on an object, the rope gets taut or stretched. A rope that is stretched is under *tension* (T). T is the amount of force that the rope exerts on the object.

Ropes (or strings, chains, cables, etc.) only pull. They never push.

1. A mass of 5 kg hangs from the ceiling by a rope. What is the tension of the rope?

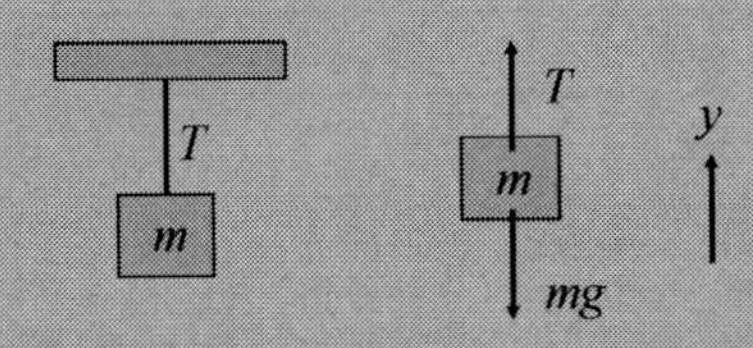

The force diagram is on the right.
From the diagram, $F_y = T - mg$.

The object is not accelerating, so $F_y = ma_y = 0$.

$0 = T - mg$
$mg = T$
$T = (5)(10) =$ **50 N**

2. A string pulls horizontally on a cat toy. The mass of the toy is 0.1 kg and its acceleration is 2 m/s². What is the tension of the string?

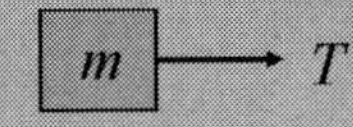

From the diagram, $F_x = T$

The toy is accelerating. $F_x = ma_x = (0.1)(2) = 0.2$ N

$T =$ **0.2 N**

Quiz 3.2

1. A child pulls a wagon in the $-x$ direction. The child pulls with a force $F_x = -5$ N. What force does the wagon exert on the child?

2. An athlete pushes a basketball upward with a force 10 N. What is the force that the basketball exerts on the athlete?

3. A wrecking ball of mass 300 kg is attached to a crane by a cable. The ball is motionless. What is the tension in the cable?

4. In a toy train, a caboose of mass 0.1 kg is pulled horizontally by a string. The acceleration is 0.5 m/s^2. What is the tension in the string?

Normal force

An object sits on a horizontal surface. The force of gravity acts downward. Why doesn't the object accelerate downward?

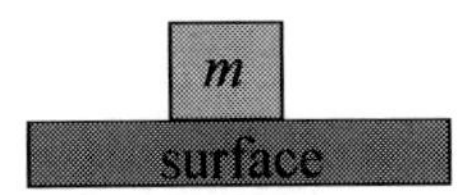

The reason is that the surface exerts a force on the object. The direction of this force is perpendicular, or *normal*, to the surface. We label this force n.

Below is a force diagram for the object. Notice that the n vector is perpendicular to the surface, and points away from the surface.

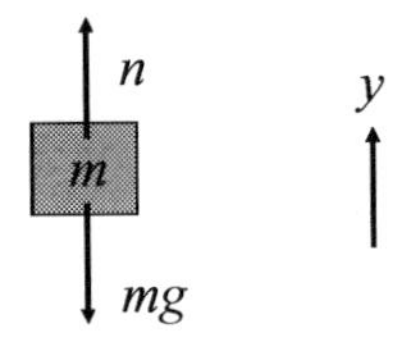

1. An object of mass 6 kg sits on a table. What is the normal force on the object?

From the diagram, $F_y = n - mg$.
The object is not accelerating, so $F_y = ma_y = 0$.

$0 = n - mg$
$mg = n$
$n = (6)(10) =$ **60 N**

2. An astronaut of mass 100 kg sits on a seat in a rocket. The rocket accelerates upward 10 m/s². What is the normal force exerted by the seat?

From the diagram, $F_y = n - mg$
The astronaut is accelerating. $F_y = ma_y = (100)(10) = 1000$ N

$1000 = n - mg$
$1000 + mg = n$
$n = 1000 + (100)(10) =$ **2000 N**

Friction

Consider a block on a table. The surface of the block is in contact with the surface of the table. These surfaces are not perfectly smooth, so it requires effort to make the block slide. The roughness of the surfaces causes a *friction* force (f). The friction force vector is parallel to the surfaces. Friction opposes sliding. It never causes sliding.

Static friction (f_s) prevents sliding motion. It occurs only when the objects are not moving relative to each other (no sliding). The maximum magnitude of static friction is

$$f_s(\text{max}) = \mu_s n$$

where μ_s is the *coefficient of static friction*. It depends on the materials (wood, metal, etc.). A large normal force n produces a large friction force. If a force exceeds $f_s(\text{max})$, the block will begin to slide. When sliding occurs, we have *kinetic friction* (f_k),

$$f_k = \mu_k n$$

where μ_k is the *coefficient of kinetic friction*.

1. A wood block (m = 3 kg) sits on a wood table. A horizontal force F pushes on the block. How much force is required to make the block slide? Let $\mu_s = 0.2$.

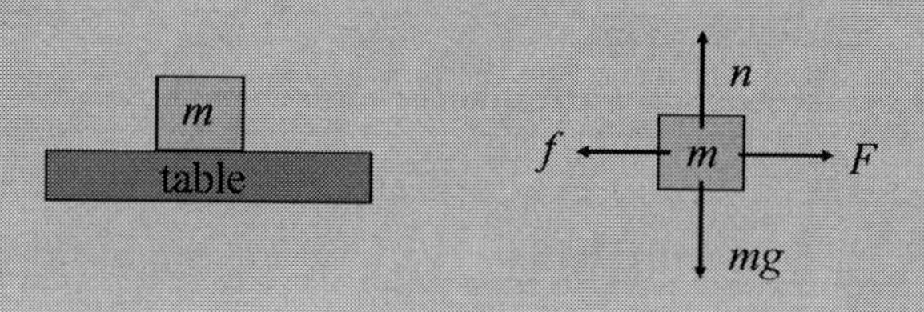

The object is not accelerating vertically, so $n = mg$
$= (3)(10) = \underline{30\text{ N}}$

$f_s(\text{max}) = (0.2)(30) = 6\text{ N}$

This is the maximum static friction force. To overcome it, $\underline{\mathbf{F > 6\text{ N}}}$.

2. In the previous problem, $F = 12$ N. Find the block's acceleration. Let $\mu_k = 0.1$.

Since 12 N exceeds $f_s(\text{max})$, the block is sliding.

$f_k = (0.1)(30) = 3\text{ N}$
$F_x = F - f_k = 12 - 3 = 9\text{ N}$

$F_x = ma_x$
$9 = (3)a_x$, so $a_x = \underline{\mathbf{3\text{ m/s}^2}}$

Quiz 3.3

1. A book has a mass of 1 kg. It sits on a shelf. What is the normal force on the book?

2. On a carnival ride, an 80 kg person sits on a chair that accelerates downward, $a_y = -5$ m/s^2. What is the normal force on the person?

3. A person pushes a cardboard box of mass 100 kg. How much force is required to make the box move? Let $\mu_s = 0.3$.

4. In the previous problem, suppose the person pushes with a horizontal force 500 N. What is the acceleration of the box? Let $\mu_k = 0.25$.

Chapter summary

Force causes acceleration	$F_x = ma_x$ $F_y = ma_y$
Gravitational force	$F_y = -mg$ (if y axis points vertically up)

- If A exerts a force on B, then B exerts an equal but opposite force on A.
- Ropes pull with a tension force T.
- The normal force n acts perpendicular to a surface and points away from the surface.
- The friction force f acts parallel to a surface and opposes sliding motion.

Static friction	$f_s(\max) = \mu_s n$
Kinetic friction	$f_k = \mu_k n$

End-of-chapter questions

1. Four forces act on an object with mass $m = 4$ kg. What is the acceleration?

1 N
1 N m 3 N
2 N

2. A large truck collides with a small car. The truck exerts a force $F_x = 10{,}300$ N on the car. What force does the car exert on the truck?

3. A 1 kg mass hangs from a string. What is the tension in the string?

4. A 0.5 kg salmon is on a plate. What is the normal force on the salmon?

5. A 1 kg mass is attached to a string. A person pulls the string upward, causing the mass to accelerate at 5 m/s^2. What is the tension in the string?

6. A horse pulls a 300 kg mass horizontally. How much force does the horse need to exert, to move the mass? The coefficient of static friction is 0.5.

7. In the previous problem, the horse pulls with 2000 N of force. What is the acceleration of the mass? The coefficient of kinetic friction is 0.3.

Challenging problems

1. A car accelerates with $a_x = 20$ m/s^2. The driver has a mass 100 kg. Ignore friction.

   a. What is the normal force of the seat bottom on the diver? (Assume the seat bottom is a horizontal surface).
   b. What is the normal force of the seat back on the driver? (Assume the seat back is a vertical surface).

2. A rope and chain are attached to an 8 kg mass. The rope pulls toward the left with a force 6 N. The chain pulls toward the right. The mass accelerates toward the right, 16 m/s^2. What is the tension in the chain? Ignore friction.

3. A person throws a 0.2 kg hockey puck onto the ice with an initial velocity 3 m/s. The coefficient of kinetic friction is 0.05. How long does it take for the puck to stop?

4. A cart accelerates toward the right. A book of mass 0.5 kg sits on the cart and does not slide. The coefficient of static friction is 0.3.

   a. Suppose the acceleration is 2 m/s^2. What is the friction force on the book?
   b. What is the maximum acceleration such that the book will not slide?

4 Forces and angles

Force vector

The previous chapter discussed horizontal and vertical forces. In this chapter, we consider forces that point along *any* direction. A *force vector* is visualized by drawing a right triangle:

F θ F_y F_x

F_x and F_y are the x- and y-components of the force vector. The magnitude of the force (F) is given by Pythagoras:

$$F = \sqrt{F_x^{\ 2} + F_y^{\ 2}}$$

From trigonometry,

$$\cos\theta = F_x / F \qquad \sin\theta = F_y / F$$

1. A force vector is given by $F_x = 4$ N, $F_y = 3$ N. Find the magnitude of the force.

$$F = \sqrt{F_x^{\ 2} + F_y^{\ 2}} = \sqrt{4^2 + 3^2} = \sqrt{25} = \underline{\mathbf{5\ N}}$$

2. A force has a magnitude 8 N. The vertical component of the force is $F_y = 4$ N. Find the horizontal angle.

$\sin\theta = 4 / 8 = 0.5$

$\theta = \sin^{-1}(0.5) = \underline{\mathbf{30°}}$

3. A force has a magnitude 50 N and a horizontal angle $\theta = 60°$. Find the horizontal component of the force vector.

$\cos\theta = F_x / F$

$\cos(60°) = F_x / 50$
$0.5 = F_x / 50$
$\underline{\mathbf{25\ N}} = F_x$

Adding force vectors

In the last chapter, we let

$$F_x = \text{the sum of forces along the } x \text{ direction}$$
$$F_y = \text{the sum of forces along the } y \text{ direction}$$

The forces and acceleration are related by

$$F_x = ma_x$$
$$F_y = ma_y$$

If we have a force that is not horizontal or vertical, we must find its x and y components. Then, we treat each component as an individual force vector. From the previous page,

$$F_x = F\cos\theta \qquad F_y = F\sin\theta$$

1. An object of mass 10 kg experiences a force of 70 N at a horizontal angle 30°. What is the acceleration?

$F_x = (70)\cos(30°) = 60.6$ N

$F_x = ma_x$
60.6 N $= (10)a_x$
6.06 m/s² $= a_x$

$F_y = (70 \text{ N})\sin(30°) = 35$ N

$F_y = ma_y$
35 N $= (10)a_y$
3.5 m/s² $= a_y$

2. Three forces act on a 1 kg mass. What is the acceleration?

Components of the 4 N vector:

$F_x = (4)\cos(70°) = 1.37$ N (+ because the vector points right)
$F_y = -(4)\sin(70°) = -3.76$ N (– because the vector points down)

$F_x = ma_x$
$-1 + 1.37 = (1)\,a_x$
0.37 m/s² $= a_x$

$F_y = ma_y$
$2 - 3.76 = (1)\,a_y$
–1.76 m/s² $= a_y$

Quiz 4.1

1. A force $F_x = 6$ N, $F_y = -6$ N is applied to an object. What is the magnitude of the force? What is the horizontal angle?

2. A force of 12 N is applied at a horizontal angle 30°. Find the horizontal component of the force vector.

3. A force acts on a 7 kg object at a horizontal angle 45°. The magnitude of the force is 10 N. What is the acceleration of the object?

4. Three forces act on a 2 kg mass. Find the acceleration.

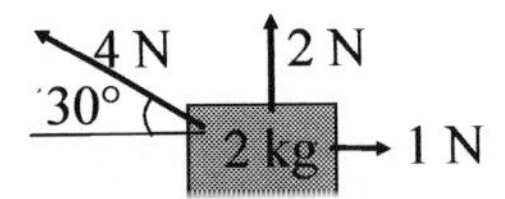

Tension

When a force acts in a direction that is not purely in the x or y direction, we must find the x and y components. If a rope pulls with a horizontal angle θ, the components are

$$F_x = T\cos\theta \qquad\qquad F_y = T\sin\theta$$

where T is the tension.

1. An object is suspended by two identical ropes. What is the tension in each rope?

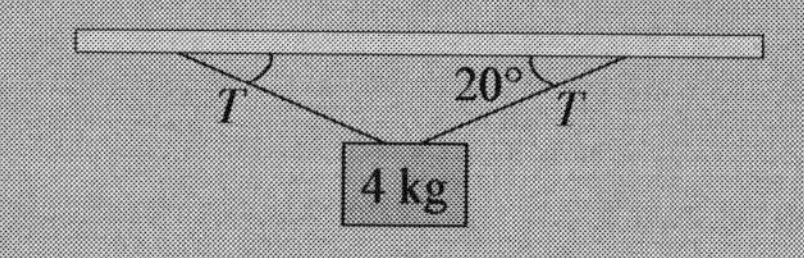

For one rope, $F_y = T\sin\theta = T\sin(20°) = 0.342\ T$

For both ropes, $F_y = 2 \times 0.342\ T = \underline{0.684\ T}$

Force of gravity: $F_y = -mg = -(4)(10) = \underline{-40\ \text{N}}$

Since the object is not accelerating, the total F_y must equal zero:

$$F_y = 0.684\ T - 40 = 0$$
$$0.684\ T = 40$$
$$T = 40/0.684 = \underline{\mathbf{58.48\ N}}$$

2. A child pulls a 10 kg sled by a rope, causing an acceleration $a_x = -2$ m/s^2. What is the tension in the rope? Ignore friction.

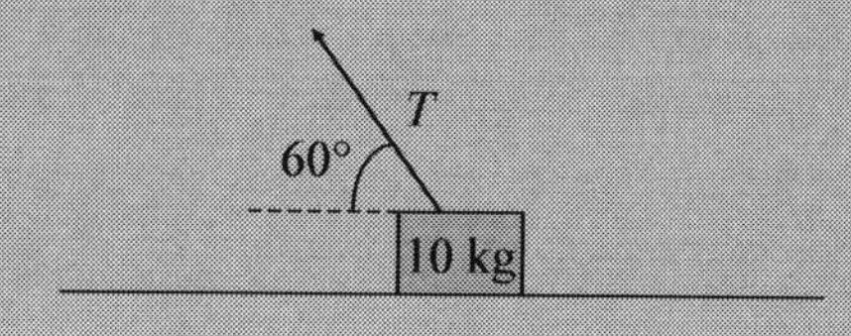

$F_x = -T\cos(60°) = \underline{-0.5\ T}$ (– because the vector points left)

$F_x = ma_x$
$-0.5\ T = (10)(-2)$
$T = 20/0.5 = \underline{\mathbf{40\ N}}$

Normal force

Here are examples of calculating the normal force, when another force acts at some angle. In these examples, the object is not accelerating in the y direction. Therefore, the sum of the forces in the y direction must be zero.

1. A person pushes a cart of mass 20 kg. The person exerts a force 10 N at a horizontal angle 55°. What is the normal force n of the ground on the cart?

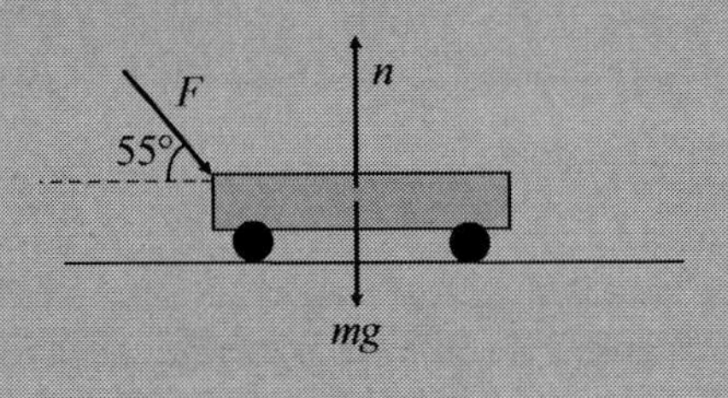

The person's force: $F_y = F \sin\theta = -(10)\sin(55°) = \underline{-8.19\text{ N}}$

Sum of all y forces: $F_y\text{ (total)} = n - mg - 8.19 = 0$

$$\begin{aligned} n &= mg + 8.19 \\ &= (20)(10) + 8.19 \\ &= \underline{\mathbf{208.19\ N}} \end{aligned}$$

2. A rope pulls a 50 kg block with a tension 200 N at a horizontal angle 30°. What is the friction force? The coefficient of kinetic friction is 0.1.

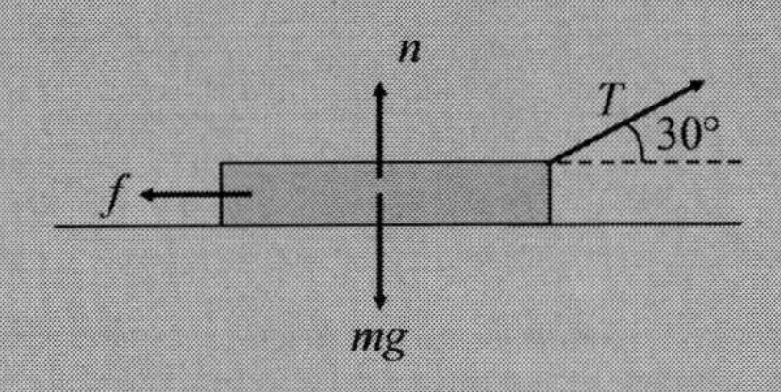

The tension force: $F_y = T\sin(30°) = (200\text{ N})(0.5) = \underline{100\text{ N}}$

Sum of all y forces:

$$\begin{aligned} F_y &= n - mg + 100 = 0 \\ n &= mg - 100 \\ &= (50)(10) - 100 \\ &= \underline{400\text{ N}} \end{aligned}$$

Friction force: $f = \mu_k n = (0.1)(400) = \underline{\mathbf{40\ N}}$

(it points in the $-x$ direction, as shown in the figure)

Quiz 4.2

1. A 0.1 kg object is suspended by two wires. Each wire is at a horizontal angle 1°. Find the tension in each wire.

2. A 2 kg object is on a frictionless floor. A person pushes on the object, at a horizontal angle 30°. The person exerts a force 4.62 N. Find the acceleration of the object.

3. In the previous problem, what is the normal force on the object?

4. A rope pulls on a block. The coefficient of kinetic friction is 0.05. Find the friction force (magnitude and direction).

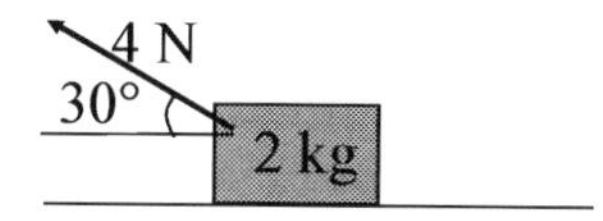

Inclined plane without friction

A block is on a frictionless plane, which is inclined at a horizontal angle θ. We choose axes where x is parallel to the surface and y is perpendicular. Imagine slowly increasing θ from zero. The axes tilt by an angle θ. Notice how the y axis makes an angle θ with vertical.

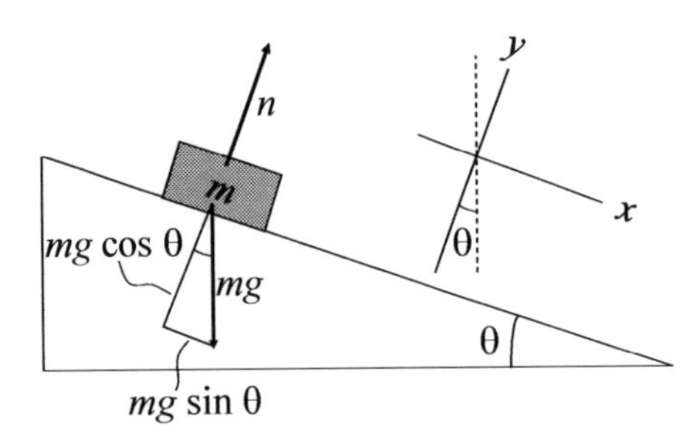

The normal force n points in the y direction. However, we must find the x and y components of mg. The x component is $mg \sin\theta$. The y component is $-mg \cos\theta$ (– because it points down).

The mass slides on the surface, accelerating in the x direction. It does *not* accelerate in the y direction, since y is perpendicular to the surface.

1. What is the acceleration of a 4 kg mass on a frictionless inclined plane with horizontal angle 30°?

$F_x = ma_x$

From the diagram, $F_x = mg \sin\theta$

$ma_x = mg \sin\theta$
$a_x = g \sin\theta$

$= (10)\sin(30°) = \underline{\mathbf{5\ m/s^2}}$

2. In the previous problem, what is the normal force?

The mass does not accelerate in the y direction, so $F_y = 0$.

$F_y = n - mg\cos\theta = 0$
$n = mg\cos\theta$

$= (4)(10)\cos(30°) = \underline{\mathbf{34.64\ N}}$

Inclined plane with friction

We now include the friction force f, which opposes sliding. Since the mass wants to slide in the $+x$ direction, friction points in the $-x$ direction.

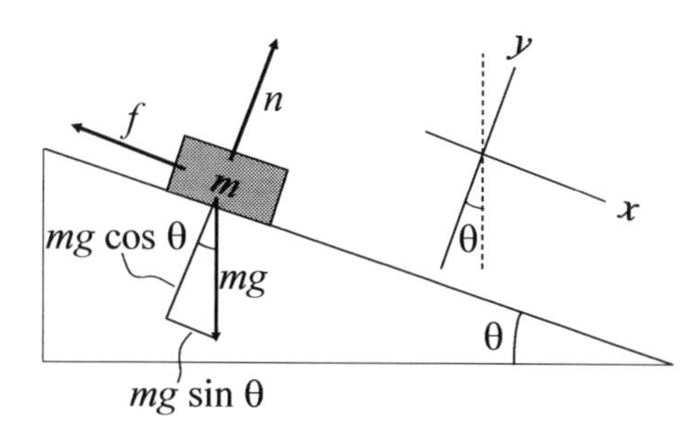

1. A block sits on an inclined plane. The angle θ is slowly increased from 0. At what angle will the block begin to slide? Let $\mu_s = 0.2$.

$n = mg \cos\theta$

$f_s\,(\text{max}) = \mu_s n = \underline{\mu_s mg \cos\theta}$

From the diagram, $F_x = mg \sin\theta - f$

The block will slide when $F_x > 0$:

$$mg \sin\theta > f$$
$$mg \sin\theta > \mu_s\, mg \cos\theta$$
$$\sin\theta > \mu_s \cos\theta$$
$$\tan\theta > \mu_s$$

To find the angle, solve $\underline{\tan\theta = \mu_s}$

$$\theta = \tan^{-1}(\mu_s) = \tan^{-1}(0.2) = \mathbf{\underline{11.3^\circ}}$$

2. In the previous problem, if θ = 30°, find the acceleration. Let $\mu_k = 0.1$.

$F_x = ma_x$

From the diagram, $F_x = mg \sin\theta - f$
$= mg \sin\theta - \mu_k\, mg \cos\theta$

$ma_x = mg \sin\theta - \mu_k\, mg \cos\theta$
$a_x = g \sin\theta - \mu_k\, g \cos\theta$

$$= (10)(0.5) - (0.1)(10)(0.866) = \mathbf{\underline{4.134\ m/s^2}}$$

Quiz 4.3

1. A mass slides down a 45° frictionless incline. What is the acceleration?

2. A 3 kg mass is on a plane inclined at a horizontal angle 20°. Find the normal force.

3. A mass sits on an inclined plane. What is the minimum angle such that the mass will slide? The coefficient of static friction is 0.4.

4. A mass slides down a 60° incline. The coefficient of kinetic friction is 0.2. Find the acceleration.

Chapter summary

Force vector

$F_x = F \cos\theta$

$F_y = F \sin\theta$

Inclined plane

- Axes are tilted by an angle θ
- x and y components of mg are shown
- n points in the y direction
- f opposes sliding

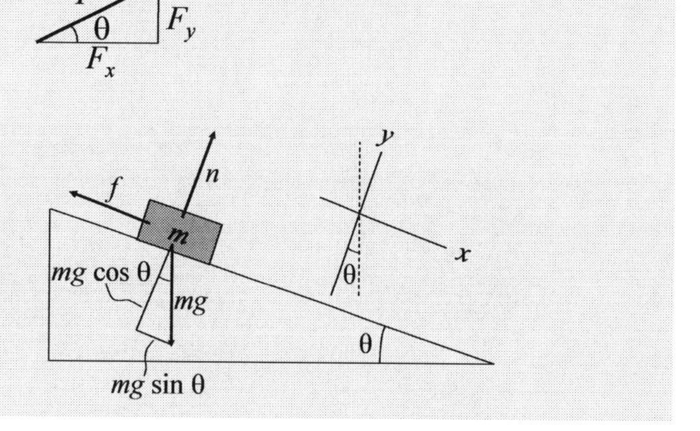

End-of-chapter questions

1. Four forces act on an object with mass $m = 4$ kg. What is the acceleration?

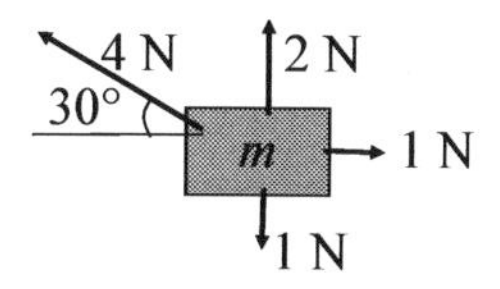

2. A mass $m = 2$ kg is suspended by two cables. What is the tension in each cable?

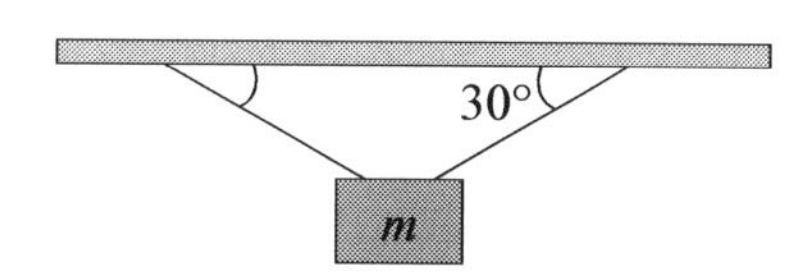

3. A person pulls a 100 kg object with 30 N of force at a horizontal angle 50°. What is the acceleration? The object is on a frictionless horizontal surface.

4. In the previous problem, what is the normal force?

5. A block slides down a frictionless plane inclined at 30°. What is the acceleration?

6. In the previous problem, what is the acceleration if the coefficient of kinetic friction is 0.25?

7. The angle of an inclined plane is increased. When the angle is 27°, a block on the plane begins to slide. What is the coefficient of static friction?

Challenging problems

1. A person pushes down on a 40 kg cart at a 53° angle, with 100 N of force. Ignore friction.

    a. What is the normal force on the cart?
    b. Assume the cart starts from rest. What is the displacement after 4 s?

2. Find the acceleration of the 10 kg mass when (a) $T = 100$ N, and (b) $T = 200$ N. Ignore friction.

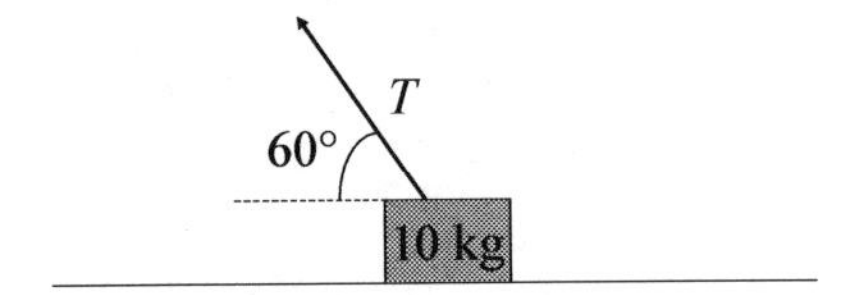

3. An 8 kg object sits on a plane that is inclined at 10°. The coefficient of static friction is 0.2. Find (a) the normal force, and (b) the friction force.

4. A worker wants to push a 100 kg crate *up* a 30° slope. The coefficient of static friction is 0.8. What is the minimum force (magnitude) the worker must exert?

5 Pulleys

Simple pulley

In a simple pulley, a rope and disc are used to redirect a force. A person pulls on the rope with a force of magnitude F. (The rope pulls on the person with an equal but opposite force.) The tension of the rope is equal to F. If the person pulls with $F = 7$ N, then $T = 7$ N.

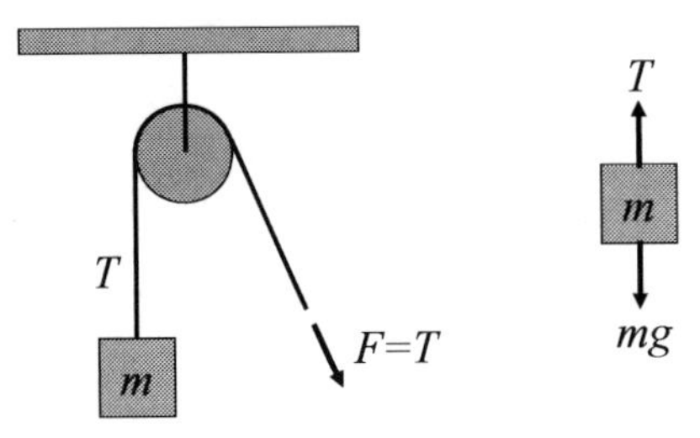

A force diagram for the mass is shown on the right. The rope pulls up while gravity acts down.

1. A 20 kg mass is connected to a simple pulley. A person pulls on the rope with 300 N of force. What is the acceleration of the mass?

From the diagram, $F_y = T - mg$

$F_y = ma_y$
$ma_y = T - mg$

$(20)a_y = 300 - (20)(10)$
$a_y = 100/20 =$ **5 m/s²**

2. In the previous problem, what is the minimum amount of force required to lift the mass?

From the diagram, $F_y = T - mg$

To lift the mass, $F_y > 0$
Or, $T > mg$
$T > (20)(10) = 200$ N

We say that the "minimum" force required is **200 N**.

This means that the tiniest amount of additional force will lift the mass. For example, 200.0000001 N will slowly lift the mass.

Compound pulley

In a compound pulley, we attach a (massless) disc to the mass. The rope winds around this disc and is attached to the ceiling. The tension T is the same along the length of the rope.

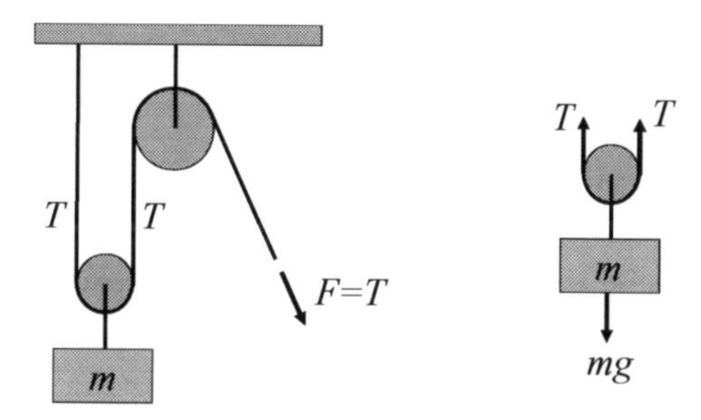

The force diagram is shown on the right. The left part of the rope, which attaches to the ceiling, pulls up with tension T. The right part of the rope, which is pulled by the person, also pulls up with tension T.

1. A 20 kg mass is connected to the compound pulley shown in the figure. A person pulls on the rope with 300 N of force. What is the acceleration of the mass?

From the diagram, $F_y = 2T - mg$

$F_y = ma_y$
$ma_y = 2T - mg$
$(20)a_y = 2(300) - (20)(10)$
$a_y = 400/20 =$ **20 m/s²**

2. In the previous problem, what is the minimum amount of force required to lift the mass?

From the diagram, $F_y = 2T - mg$

To lift the mass, $F_y > 0$
Or, $2T > mg$
$T > (20)(10)/2 = 100$ N

The minimum force required is **100 N**

Quiz 5.1

1. A 3 kg mass is attached to a simple pulley. A wimpy person exerts a force of only 6 N on the rope. What is the acceleration of the mass?

2. In the previous problem, what is the minimum force required to lift the mass?

3. A bucket of water of mass 50 kg is connected to a compound pulley. A machine pulls on the rope with a force of 900 N. What is the acceleration of the bucket?

4. In the previous problem, what is the minimum force required to lift the bucket?

Atwood's machine

In Atwood's machine, two masses are connected to each other via a simple pulley. The lighter mass m accelerates up, with $a_y = a$. The heavier mass M accelerates *down*, with $a_y = -a$.

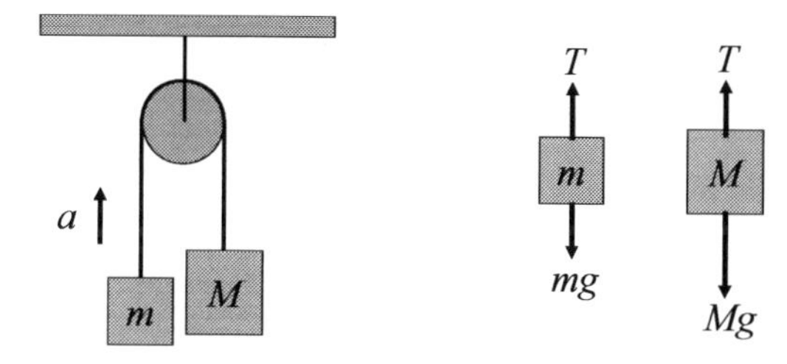

The force diagrams are shown on the right. To find the acceleration, first solve for T for each mass. Then set the 2 equations equal to each other and solve for a.

1. An Atwood's machine has masses $m = 1$ kg and $M = 3$ kg. Find the acceleration of mass m.

From the m diagram, $F_y = T - mg$

$F_y = ma$

$T - mg = ma$

$T = mg + ma \quad = \quad \underline{10 + (1)a}$

From the M diagram, $F_y = T - Mg$

$F_y = M(-a)$ (– because it accelerates down)

$T - Mg = -Ma$

$T = Mg - Ma \quad = \quad \underline{30 - (3)a}$

Set the 2 underlined equations equal: $10 + a = 30 - 3a$

$4a = 20$

$a = \underline{\mathbf{5\ m/s^2}}$

2. In the previous problem, what is the tension in the rope?

From the first underlined equation, $T = 10 + (1)a$

$= 10 + 5 = \underline{\mathbf{15\ N}}$

Pulley with vertical and horizontal masses

In this example, two masses are connected via a pulley. One mass (m) is hanging. The other mass (M) is on a horizontal surface. The mass m accelerates down while M accelerates toward the left.

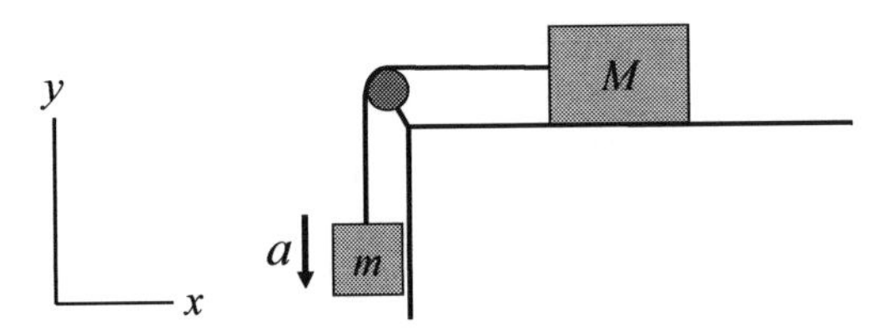

To find the acceleration, first solve for T for each mass. Then set the 2 equations equal to each other and solve for a. Ignore friction.

1. In the figure, $m = 1$ kg and $M = 3$ kg. Find the acceleration of mass m.

For mass m,

$F_y = T - mg$

$F_y = m(-a)$ (– because it accelerates down)

$T - mg = -ma$

$T = mg - ma$ = $\underline{10 - (1)a}$

For mass M,

$F_x = -T$ (– because T pulls to the left)

$F_x = M(-a)$ (– because M accelerates left)

$T = Ma$ = $\underline{3a}$

Set the 2 underlined equations equal:

$10 - a = 3a$

$10 = 4a$

2.5 m/s² $= a$

(in the $-y$ direction)

2. In the previous problem, what is the tension in the rope?

From the first underlined equation, $T = 10 - (1)a$

$= 10 - 2.5 =$ **7.5 N**

Quiz 5.2

1. An Atwood's machine has masses 10 kg and 40 kg. What is the acceleration of the heavier mass?

2. In the previous problem, what is the tension in the rope?

3. Two masses, $m = 3$ kg and $M = 7$ kg, are connected via a pulley. What is the acceleration of the heavier mass? Ignore friction.

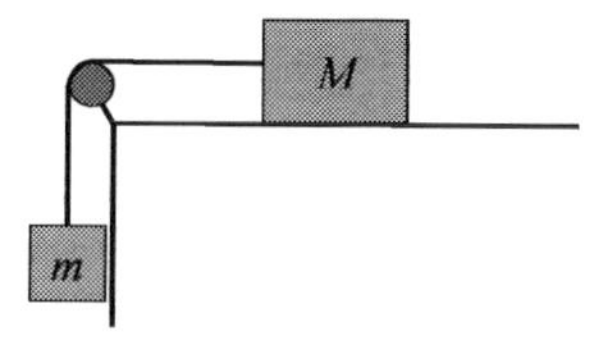

4. In the previous problem, what is the tension in the rope?

Pulley and friction

Let's look at the previous problem, but add the friction force. The force diagram for M is shown on the right.

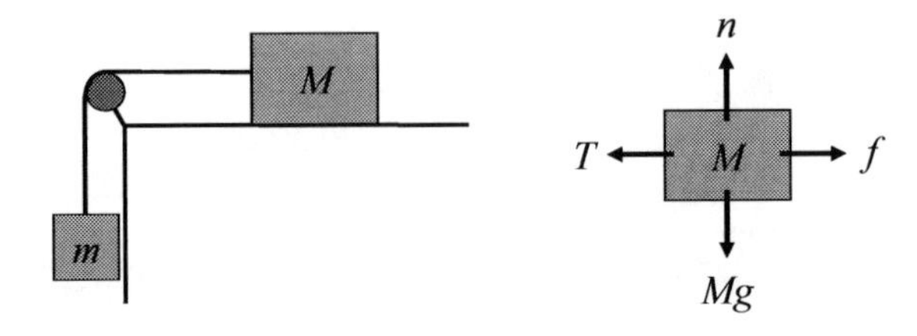

1. In the figure, $M = 3$ kg and $\mu_s = 0.4$. Because of friction, the masses are not moving. What is the maximum value of m such that the masses will not move?

For mass m,

$$F_y = T - mg$$
$$F_y = 0 \quad \text{(the mass is not accelerating)}$$

$$T - mg = 0$$
$$\underline{T = mg = m(10)}$$

For mass M,

$$F_x = -T + f$$
$$= -T + \mu_s Mg$$
$$F_x = 0 \quad \text{(the mass is not accelerating)}$$

$$-T + \mu_s Mg = 0$$
$$-m(10) + (0.4)(3)(10) = 0$$
$$\underline{\mathbf{1.2\ kg}} = m$$

If m is greater than this, the static friction force will be overcome and the masses will accelerate.

Pulley and inclined plane

Two masses are connected via a pulley. Mass m is hanging vertically. Mass M is on an inclined plane. We use different axes for each mass. Ignore friction.

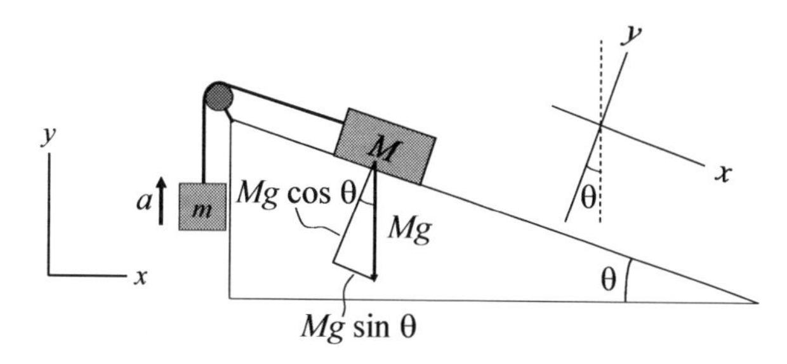

1. In the figure, $M = 5$ kg, $m = 3$ kg, and $\theta = 30°$. Find the acceleration of m.

For mass m,

$$F_y = T - mg$$
$$F_y = ma$$
$$T - mg = ma$$
$$T = mg + ma \quad = \quad \underline{30 + 3a}$$

For mass M,

$$F_x = -T + Mg \sin\theta$$
$$F_x = Ma$$
$$-T + Mg \sin\theta = Ma$$
$$T = Mg \sin\theta - Ma \quad = \quad \underline{25 - 5a}$$

Set the 2 underlined equations equal:

$$30 + 3a = 25 - 5a$$
$$8a = -5$$
$$\underline{\mathbf{a = -0.625\ m/s^2}}$$

In the figure, a points up. However, we found a negative value for a. That means that m accelerates *down*.

Quiz 5.3

1. Two masses are connected via a pulley, with $m = 5$ kg. What is the minimum mass M such that it will not slide? The coefficient of static friction is 0.2.

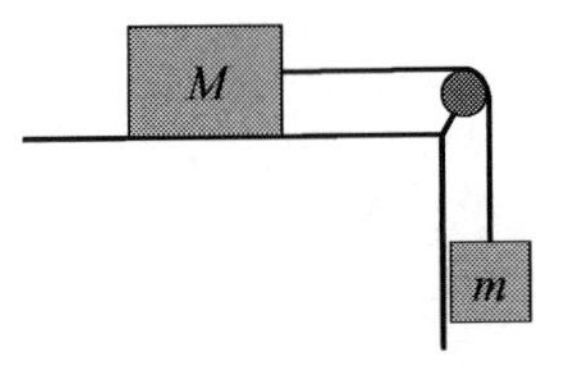

2. In the previous problem, what is the tension in the rope?

3. Two masses are connected via a pulley. A mass ($m = 5$ kg) hangs vertically, while the other mass ($M = 20$ kg) is on a frictionless inclined plane, with $\theta = 30°$. Find the acceleration of m.

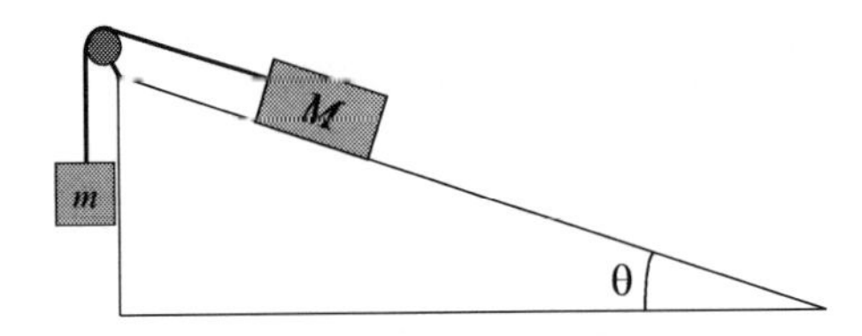

4. In the previous problem, what is the tension in the rope?

Chapter summary

- Pulleys re-direct the tension force
- A rope pulls on an object with tension T

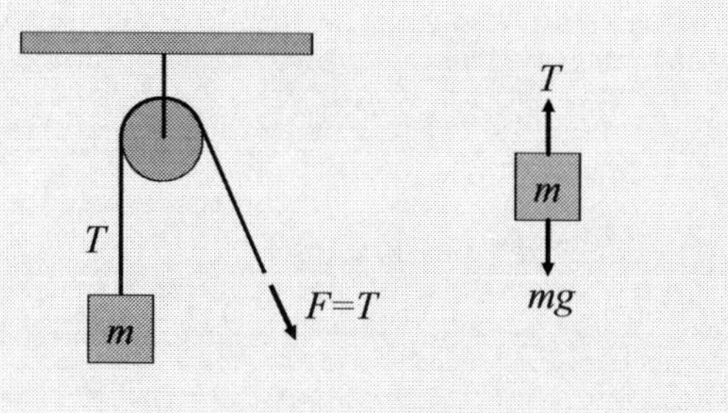

End-of-chapter questions

1. A person wants to lift a 5 kg mass with a pulley. What is the minimum force that the person must exert, for (a) a simple pulley, and (b) a compound pulley?

2. An Atwood's machine has two masses that are 5 kg each.

   a. What is the tension in the rope?
   b. An additional 5 kg is added to one of the masses. What is the acceleration of the lighter mass?

3. Two masses, $m = 5$ kg and $M = 15$ kg, are connected via a pulley. What is the acceleration of the hanging mass? Ignore friction.

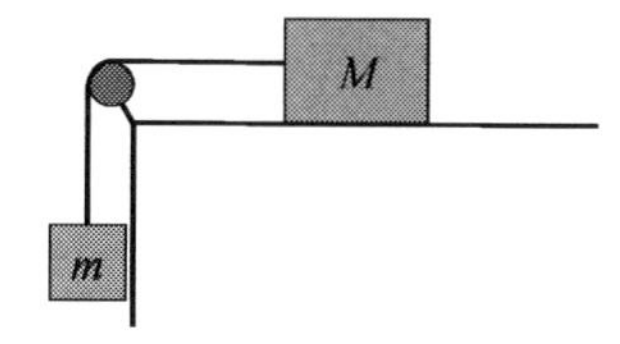

4. In the previous problem, let $\mu_s = 0.4$.

   a. What is the friction force on M?
   b. What is the maximum value that m can be and still not move?

5. Two equal masses are connected via a pulley. One mass hangs vertically and the other mass is on a frictionless inclined plane, with $\theta = 32°$. Find the acceleration of the hanging mass.

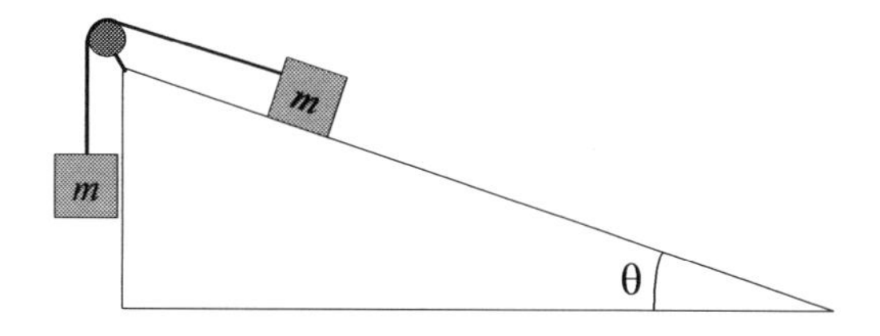

Challenging problems

1. Two people are lifting a 100 kg object with a compound pulley. One person exerts a force 400 N on the rope. The second person pushes up on the object directly, with a force 500 N. What is the acceleration of the object?

2. An Atwood's machine has masses 10 kg and 5 kg. Initially, they are 15 m above the floor. How long does it take for the 10 kg mass to hit the floor?

3. Two masses, m = 5 kg and M = 15 kg, are connected via a pulley. For M, the coefficient of kinetic friction is 0.1. The masses are accelerating. What is the acceleration of M?

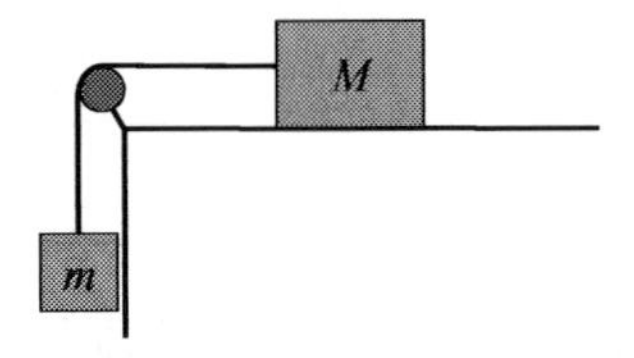

4. Two masses are connected via a pulley. A mass (m = 5 kg) hangs vertically, while the other mass (M = 20 kg) is on an inclined plane, with $\theta = 30°$. For M, the coefficient of kinetic friction is 0.1. The masses are accelerating. Find the acceleration of M.

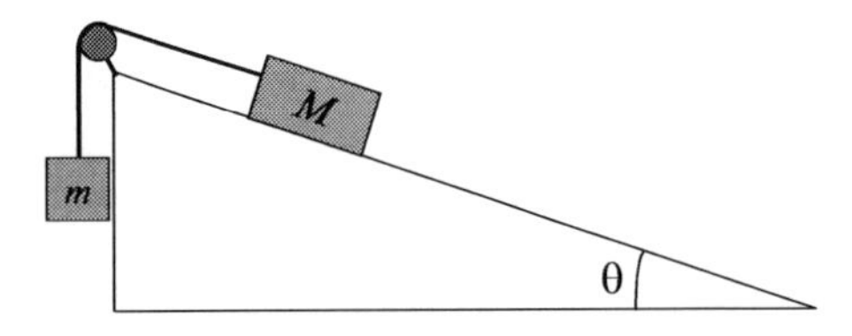

6 Circular motion

Centripetal acceleration

In *uniform circular motion*, an object travels in a circle of radius R, with a constant speed v. The velocity is *tangential* to the circle. Even though the speed doesn't change, the *direction* of the velocity vector continuously changes as the object goes around. This is a type of acceleration, since acceleration is change in velocity over time. We call it *centripetal acceleration*, where "centripetal" means pointing toward the center.

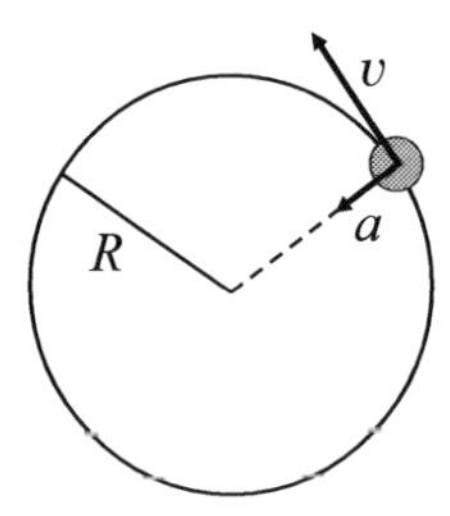

Follow these rules to calculate centripetal acceleration:

- The acceleration vector points toward the center of the circle.
- The magnitude of the acceleration is $a = v^2/R$.

1. An object travels in a circular path with a speed 5 m/s. The radius of the circle is 10 m. Find the acceleration.

$a = v^2/R$
$= 25/10 =$ **<u>2.5 m/s²</u>** (Direction: toward center of circle)

2. A bug is on a disc that rotates at 77 rotations per minute (rpm). The distance from the bug to the disc center is 10 cm. What is the bug's acceleration?

Circumference $= 2\pi R = 2(3.14)(0.1 \text{ m}) = 0.63 \text{ m}$

$$v = \frac{77 \text{ rotations}}{\text{min}} \frac{1 \text{ min}}{60 \text{ s}} \frac{0.63 \text{ m}}{\text{rotation}} = 0.81 \text{ m/s}$$

$a = v^2/R = (0.81^2)/0.1 =$ **<u>6.6 m/s²</u>** (Direction: toward center of circle)

Centripetal tension force

To keep an object on a circular path, there must be a centripetal force. Consider an object attached to a string, on a frictionless table. The object undergoes uniform circular motion. We define the y direction to point from the center to the object.

(Top view)

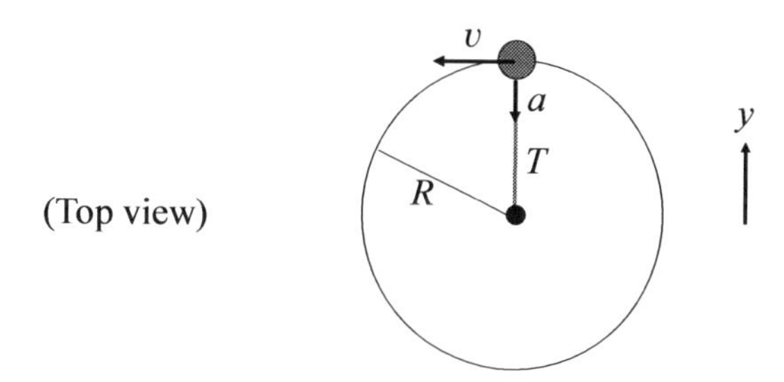

The string tension T is the only force along the y direction. It pulls the ball toward the center, and is referred to as a *centripetal force*. (If someone cuts the string, the ball will go in a straight line.)

1. A 0.2 kg ball attached to a string is on a frictionless table. The ball travels in a horizontal circular path with a speed 5 m/s. The radius of the circle is 0.1 m. Find the tension of the string.

$a_y = -v^2/R$ (– because a points toward center)
$F_y = -T$ (– because T points toward center)
$F_y = ma_y$
$T = mv^2/R$ (canceling the minus signs)

$= (0.2)(25)/0.1 =$ **50 N**

2. In the previous problem, what is the normal force of the table on the ball?

(Side view)

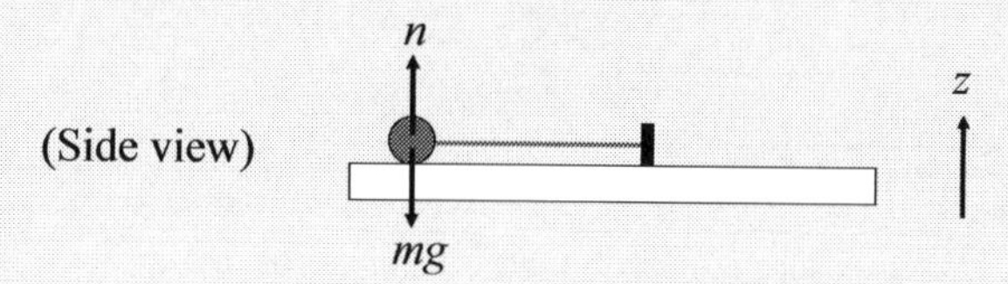

Define z to point up.
$F_z = n - mg$
Since the ball does not accelerate in the z direction, $F_z = 0$

$n = mg$
$= (0.2)(10) =$ **2 N**

Quiz 6.1

1. An object travels in a circular path with a speed 3 m/s. The radius of the circle is 9 m. Find the acceleration.

2. A small rock is stuck on a car tire. The tire has a diameter of 75 cm and spins at 1200 rpm. What is the acceleration of the rock?

3. A 1 kg mass is attached to a rope on a horizontal frictionless surface. The mass travels in a horizontal circle of radius 0.5 m, with a constant speed 8 m/s. Calculate the tension of the rope.

4. In the previous problem, what is the normal force on the mass?

Centripetal friction force

A car travelling around a curve requires the centripetal force of friction. Otherwise, it will travel in a straight line and go off the road. This can happen when the road is icy.

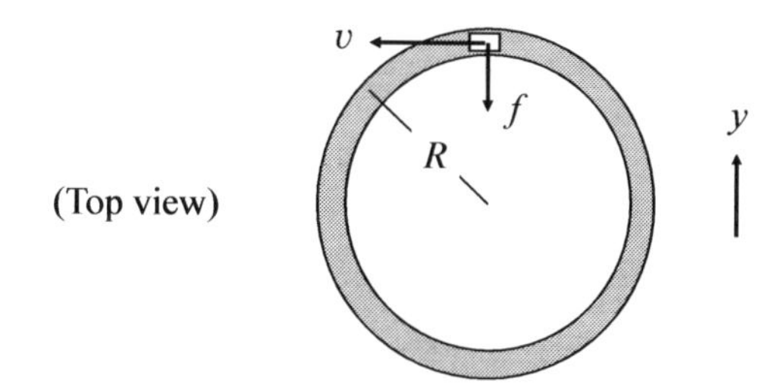

In the figure, a car is going around a circular horizontal track at a speed v. The friction force f acts in the $-y$ direction. Since f prevents sliding, it is considered static friction, even though the car is moving.

1. A 1500 kg car travels around a horizontal circular track of radius 500 m at a speed 20 m/s. What is the centripetal friction force?

$a_y = -v^2/R$
$F_y = -f$

$F_y = ma_y$
$\underline{f = mv^2/R}$

$= (1500)(400)/500 = \underline{\mathbf{1200\ N}}$

2. In the previous problem, the coefficient of static friction is 0.5. What is the maximum speed the car can travel without sliding?

$f(\text{max}) = \mu_s mg$
$f(\text{max}) = mv^2/R$

$\mu_s mg = mv^2/R$
$\mu_s g = v^2/R$
$\mu_s gR = v^2$

$v^2 = (0.5)(10)(500) = 2500$
$v = \sqrt{2500} = \underline{\mathbf{50\ m/s}}$

Centripetal normal force

The normal force can act as a centripetal force. For example, there is a carnival ride that spins people around a horizontal circle. The wall exerts a normal force (n) on a person. This force is necessary to keep the person moving in a circle.

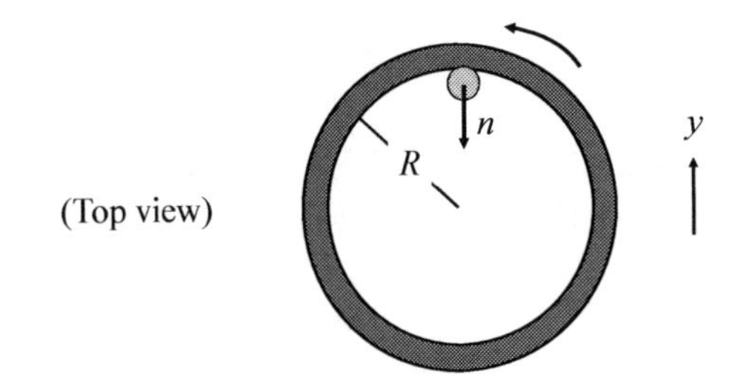

1. An 80 kg person stands against the wall in a rotating cylinder. The radius of the cylinder is 10 m and it spins at 60 rpm. What is the normal force?

$a_y = -v^2/R$
$F_y = -n$

$F_y = ma_y$
$\underline{n = mv^2/R}$

Circumference $= 2\pi R = 2(3.14)(10\text{ m}) = 63\text{ m}$

$$v = \frac{60\text{ rotations}}{\text{min}}\frac{1\text{ min}}{60\text{ s}}\frac{63\text{ m}}{\text{rotation}} = 63\text{ m/s}$$

$n = (80)(63^2)/10 =$ **31,800 N**

2. When the cylinder is spinning, it can hold the person up with only the friction force. In the previous problem, what coefficient of static friction (between the person and the wall) is required to hold the person up?

(Side view)

$f(\text{max}) = \mu_s n = mg$

$\mu_s = mg/n = (80)(10)/(31{,}800) =$ **0.025**

Quiz 6.2

1. A 100 kg go-cart travels around a horizontal circular track of radius 50 m at a speed 5 m/s. The coefficient of static friction is 0.2. What is the centripetal friction force?

2. In the previous problem, what is the maximum speed that the go-cart can go and not slide?

3. A cylinder has a diameter 0.32 m and stands vertically. It spins at 120 rpm. A 0.01 kg rock is in contact with the vertical wall of the cylinder. What is the normal force on the rock?

4. In the previous problem, find the minimum coefficient of friction such that the rock will stay in contact with the wall.

Motion in a vertical circle

The previous examples involved motion in a horizontal circle. Now, suppose a person holds onto a ball and string, and makes the ball travel in a vertical circle. Defining z to point up, gravity acts in the $-z$ direction.

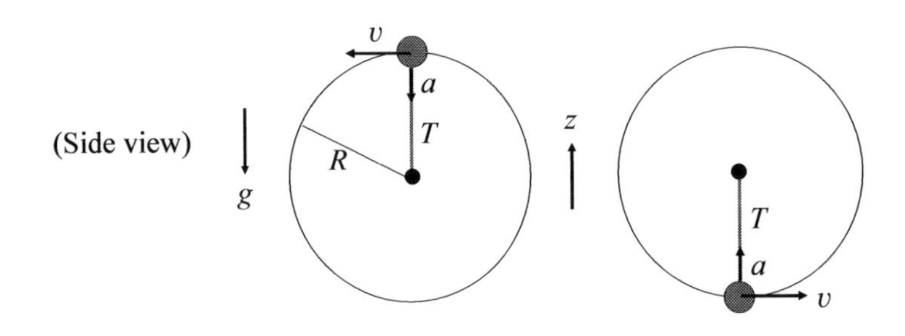

1. A 0.2 kg ball undergoes motion in a vertical circle of radius $R = 0.5$ m with a speed $v =$ 3 m/s. What is the tension of the string when the ball is at the top of the circle?

$a_z = -v^2/R$
$F_z = -T - mg$

$F_z = ma_z$
$\underline{T + mg = mv^2/R}$

$T + (0.2)(10) = (0.2)(9)/0.5$
$T + 2 = 3.6$
$T = \underline{\mathbf{1.6\ N}}$

2. In the previous problem, what is the tension when the ball is at the bottom of the circle?

$a_z = v^2/R$ (+ because acceleration points up)
$F_z = T - mg$ (T also points up)

$F_z = ma_z$
$\underline{T - mg = mv^2/R}$

$T - (0.2)(10) = (0.2)(9)/0.5$
$T - 2 = 3.6$
$T = \underline{\mathbf{5.6\ N}}$

Bucket of water

In this example, a person swings a bucket of water in a vertical circle. If the speed is high enough, the water will be in contact with the bucket, which exerts a normal force (n). If the speed is too low, then the water will lose contact with the bucket. Then, $n = 0$ and the person will get wet!

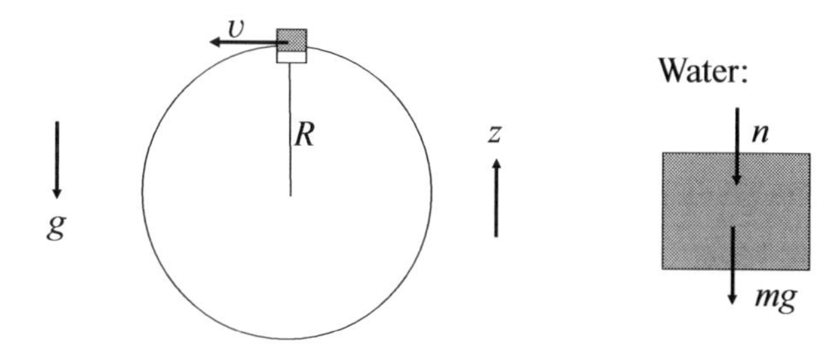

1. A person swings a bucket of water in a circle of radius 0.5 m and speed 3 m/s. The water has a mass 4 kg. Find the normal force of the bucket on the water, at the top of the circle.

$a_z = -v^2/R$
$F_z = -n - mg$

$F_z = ma_z$
$n + mg = mv^2/R$
$\underline{n = mv^2/R - mg}$

$n = (4)(9)/0.5 - (4)(10) =$ **<u>32 N</u>**

2. In the previous problem, what is the minimum speed such that the water remains in contact with the bucket?

$n = mv^2/R - mg = 0$ when the water loses contact with the bucket

$$mv^2/R - mg = 0$$
$$(4)v^2/(0.5) - (4)(10) = 0$$
$$8v^2 = 40$$
$$v^2 = 5$$
$$v = \sqrt{5} = \underline{\mathbf{2.24\ m/s}}$$

Quiz 6.3

1. A 3 kg mass is held by a rope. A person makes it travel in a vertical circle of diameter 2 m at a constant speed 4 m/s. What is the tension of the rope when the ball is at the top of the circle?

2. In the previous problem, what is the tension when the mass is at the bottom of the circle?

3. A 70 kg person sits in a roller coaster car. The car goes in a vertical circle of radius 25 m. At the top of the circle, the speed is 20 m/s. What is the normal force on the person?

4. In the previous problem, what is the minimum speed such that the person remains in contact with the car?

Chapter summary

Centripetal acceleration

$a = v^2/R$

Acceleration points toward center of circle

Centripetal force

$F = ma$

Centripetal force points toward center of circle

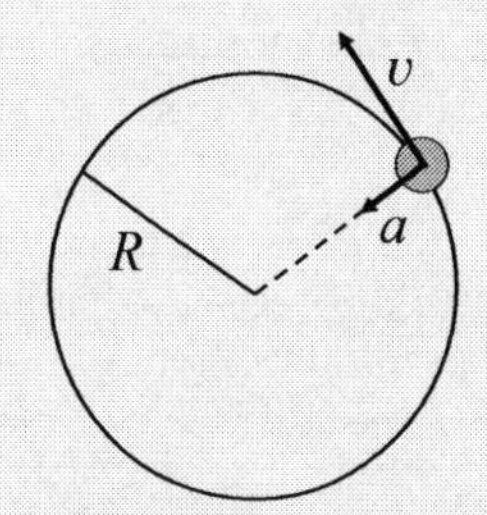

End-of-chapter questions

1. The earth is about 150,000,000 km from the sun. It takes about 365 days $\approx 3.14 \times 10^7$ s to complete one orbit. Calculate the centripetal acceleration.

2. On the space station, an astronaut holds on to a 0.3 m long string attached to a 0.05 kg mass. The mass travels in a circle at a speed 0.1 m/s. Find the tension of the string.

3. A 1 gram particle is on a disc that rotates at 120 rotations per minute (rpm). The distance from the particle to the disc center is 10 cm. What is the centripetal friction force on the particle?

4. A car travels around a circular track of radius 300 m. The coefficient of static friction is 0.4. What is the maximum speed the car can travel without sliding?

5. A bucket is filled with 3 kg water and a 0.9 m rope is attached. The bucket has a mass 1 kg. A person swings the bucket of water in a vertical circle with a speed 4 m/s.

   a. What is the tension in the rope, at the top of the circle?
   b. What is the tension at the bottom of the circle?

6. In the previous problem, what is the minimum speed such that the water remains in contact with the bucket?

Challenging problems

1. In a science fiction movie, a circular space station of radius 500 m rotates. A person standing on the edge of the space station experiences a centripetal normal force (n). Find the speed of rotation (v) that is required to simulate Earth's gravity ($n = mg$).

2. A toy of mass 0.64 kg is attached to a string of length $R = 0.32$ m. The toy is on a horizontal table and travels in a circle, at 60 rpm. The coefficient of kinetic friction is 0.05.

   a. What is the tension in the string?
   b. To keep the speed constant, the toy has a small motor. What is the force provided by the motor?

3. A 5 kg tether ball is held by a rope, at a vertical angle 45°. The ball travels in a horizontal circle of radius 1.6 m at a speed 4 m/s. What is the tension in the rope?

4. A toy mouse of mass 0.01 kg is attached to a 5 cm long string. A cat owner lets the toy swing in a vertical circular arc. At the bottom of the swing, its speed is 0.5 m/s and it is 10 cm above the floor.

   a. What is the tension of the string?
   b. The string breaks. How long does it take for the toy to hit the floor?

7 Work and energy

Work

You do positive work when you push or pull an object and it moves in the same direction as the force you exert. If you push on an elephant and it doesn't move, you haven't done any work. Work (W) is measured in Joules (J). The work done by a force is

$$W = F_{//}d$$

where $F_{//}$ is the component of the force *along the direction of motion* and d is the distance traveled. If the force is opposite to the direction of motion, then $F_{//}$ is negative.

1. A person pushes a cart with $F = 5$ N of force at a horizontal angle $\theta = 30°$, over a distance $d = 10$ m. How much work did the person do?

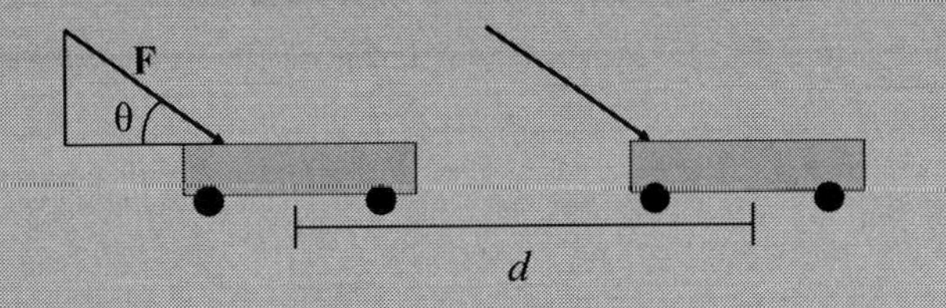

$W = F_{//}d$
$= (F \cos\theta)d$ (because the horizontal component of **F** is $F \cos\theta$)
$= (5 \cos 30°)(10) =$ **43.3 J**

2. A barbell has two masses $m = 25$ kg connected by a massless bar. A person slowly lifts the barbell a distance $h = 1$ m. How much work did the person do?

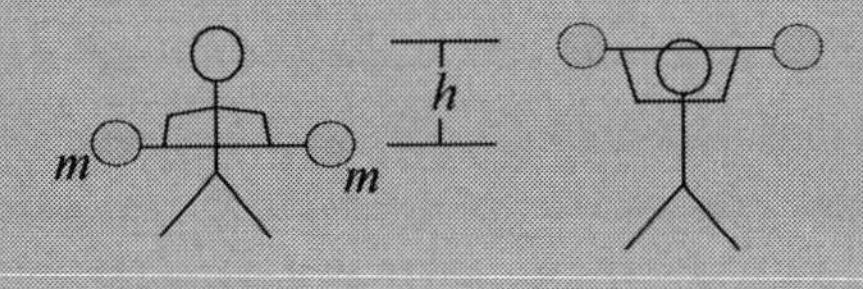

$W = F_{//}d$
$= (2mg)h$ (because the person must exert a force $2mg$)
$=$ **500 J**

3. In the previous problem, how much work did the gravity force do?

$W = (-2mg)h$ (– because gravity exerts a force opposite to the motion)
$=$ **–500 J**

Work on an inclined plane

Recall the inclined plane from Chapter 4:

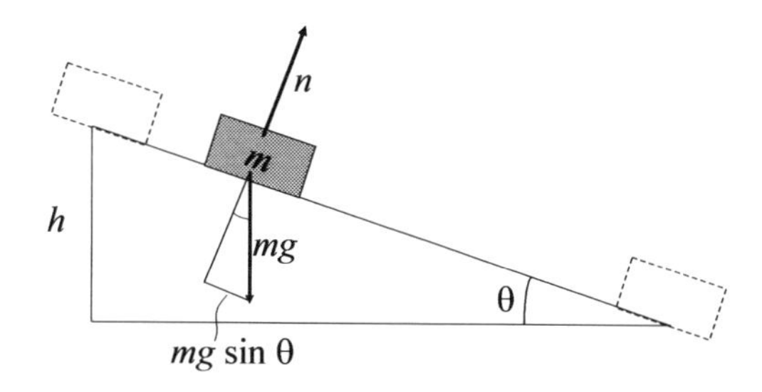

An object is initially at a height h. It slides down the inclined plane and reaches the bottom. The distance traveled, d, is the hypotenuse of the triangle. From trigonometry,

$$\sin\theta = h/d$$

1. How much work was done by the normal force?

$W = F_{//}d$

Since the normal force is perpendicular to the direction of motion, $F_{//} = 0$

So, $\underline{\boldsymbol{W = 0}}$

2. How much work was done by the gravity force?

$W = F_{//}d$

$= (mg \sin\theta)d$

$= (mg\, h/d)d = \underline{\boldsymbol{mgh}}$

Note that the work done by gravity does not depend on the angle θ.

In the following sections, we will solve this type of problem using energy conservation.

Quiz 7.1

1. A child pulls a wagon with a force of 7 N at a horizontal angle $\theta = 60°$, over a distance of 4 m. How much work did the child do?

2. A person lifts a 50 kg object 0.5 m. How much work did the person do?

3. In the previous problem, how much work was done by gravity?

4. A 1000 kg car is on a hill that is 50 m high. It rolls to the bottom of the hill. How much work was done by gravity?

Kinetic energy

Kinetic energy (K) is due to an object's motion and its units are Joules (J). It is defined

$$K = \tfrac{1}{2}mv^2$$

where m is the object's mass and v is its speed. Positive work increases K. Negative work decreases K. This change in K can be written

$$\text{Final } K = \text{Initial } K + W$$

1. A 2 kg object moves at 3 m/s at a horizontal angle of 28°. What is its kinetic energy?

$K = \tfrac{1}{2}mv^2$
$= \tfrac{1}{2}(2)(3^2) = \underline{\textbf{9 J}}$

2. A 2 kg block's initial speed is 3 m/s. A force of 8 N pushes the block over a distance of 2 m. What is the final speed?

Initial $K = 9$ J (see above)
$W = F_{//}d = (8)(2) = 16$ J
Final $K = 9 \text{ J} + 16 \text{ J} = 25$ J

$K = \tfrac{1}{2}mv^2$
$25 = \tfrac{1}{2}(2)v^2$
$25 = v^2$, so $v = \underline{\textbf{5 m/s}}$

3. A 1000 kg car travels at 20 m/s in the $-x$ direction. The driver presses the brakes and the car stops. How much work was done on the car?

Initial $K = \tfrac{1}{2}mv^2$
$= \tfrac{1}{2}(1000)(400) = 200{,}000$ J
Final $K = 0$

So, K decreased by 200,000 J

$W = \underline{\textbf{–200,000 J}}$

Potential energy

The potential energy (U) due to gravity is

$$U = mgy$$

where m is the object's mass, $g \approx 10$ m/s^2, and y is the object's height. If there is no friction or other external force, then the total energy E is constant:

$$E = K + U = \text{constant}$$

This property is called *energy conservation.*

1. A person skis down a frictionless slope, 2000 m high, starting from rest. What is the person's speed at the bottom of the slope?

Initial $E = \frac{1}{2}m0^2 + mgh = mgh$

Final $E = \frac{1}{2}mv^2 + mg0 = \frac{1}{2}mv^2$

Set them equal: $mgh = \frac{1}{2}mv^2$

$$2gh = v^2$$

$$v^2 = 2(10)(2000) = 40{,}000$$

$$\underline{v = 200 \text{ m/s}}$$

2. A roller coaster starts at a height h from rest. It rolls down and goes on a loop of radius R. What is the speed at the top of the loop?

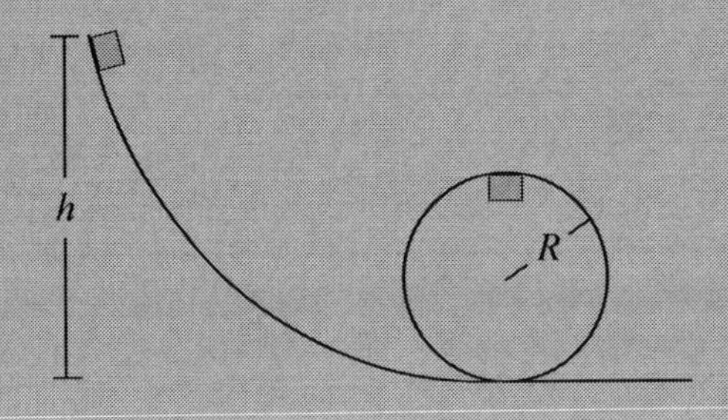

Initial $E = \frac{1}{2}m0^2 + mgh = mgh$

Final $E = \frac{1}{2}mv^2 + mg(2R)$

Set them equal: $mgh = \frac{1}{2}mv^2 + mg(2R)$

$$gh = \frac{1}{2}v^2 + 2gR$$

$$2gh - 4gR = v^2$$

$$\underline{v = \sqrt{2gh - 4gR}}$$

Quiz 7.2

1. A 4 kg cart travels on a track at 6 m/s. A person pushes the cart along the direction of motion, with a force of 8 N, over a distance of 7 m. How fast does the cart travel now?

2. A 10 kg ball travels toward a wall with a speed of 3 m/s. It bounces off the wall and travels at a speed of 1 m/s. How much work was done on the ball?

3. A penguin slides down a frictionless hill that is 20 m high. The penguin starts from rest. What is its speed at the bottom of the hill?

4. In the previous problem, the penguin continues sliding. It goes up a small hill of height 15 m. What is the penguin's speed at the top of the small hill?

Friction and thermal energy

If there is kinetic friction, mechanical energy is lost. The change in energy is

$$\text{Final } E = \text{Initial } E + W$$

Since friction opposes motion, W is negative. The lost energy is not really gone, though. It's actually transformed to thermal energy, or heat.

1. A 50 kg child slides down a 5 m high hill on a sled, with an initial speed 1 m/s. At the bottom of the hill, the child's speed is 8 m/s. How much heat was generated?

Initial $E = ½mv^2 + mgh$
$= ½(50)(1) + (50)(10)(5) = \underline{2525 \text{ J}}$

Final $E = ½mv^2 + mg0$
$= ½(50)(64) = \underline{1600 \text{ J}}$

The mechanical energy lost was 2525 – 1600 J = **925 J**
This is the amount of heat that was generated.

2. A block slides down a 20° inclined plane, starting from rest at a height $h = 3$ m. The coefficient of kinetic friction is $\mu_k = 0.04$. Find the speed at the bottom.

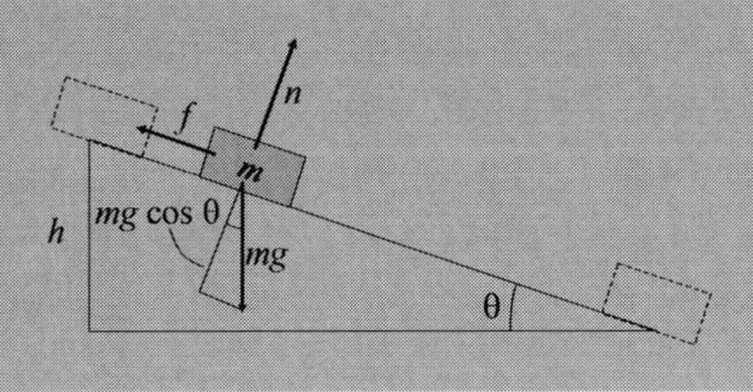

The distance traveled, d, is the hypotenuse of the triangle.
$\sin\theta = h/d$
$d = h/\sin\theta = 3/\sin(20°) = \underline{8.8 \text{ m}}$

Work due to friction:
$W = -f\,d = -(\mu_k\, mg \cos\theta)\, d$

Initial $E = mgh$
Final $E = ½mv^2$
Final E = Initial $E + W$

$½mv^2 = mgh - (\mu_k\, mg \cos\theta)\, d$
$v^2 = 2gh - 2(\mu_k\, g \cos\theta)\, d$
$= (2)(10)(3) - (2)(0.04)(10)(0.94)(8.8) = 53.4$

$\underline{\mathbf{v = 7.3 \text{ m/s}}}$

Power

Power (P) is the rate at which work is done. It is measured in units of J/s, also called Watts (W). If a force does an amount of work W in a time Δt, then its power output is

$$P = W/\Delta t$$

Recall that $W = F_{//}d$, where $F_{//}$ is the component of the force along the direction of motion. Plugging this into the above equation yields

$$P = F_{//}v$$

where v is the object's speed (m/s).

1. A person pushes a box with a force of 5 N, causing it to move at 3 m/s. What is the person's power output?

$P = F_{//}v$
$\quad = (5)(3) = \underline{\textbf{15 W}}$

2. A construction worker pushes a 50 kg crate at a constant speed of 1.5 m/s. The coefficient of kinetic friction is $\mu_k = 0.2$. Find the power output of the construction worker.

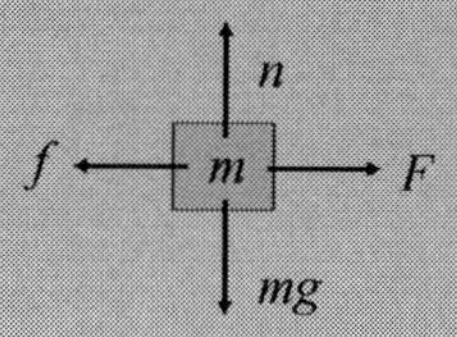

Force due to friction: $f = \mu_k mg$

Since the crate travels at a constant speed (no acceleration), the sum of the forces on the crate is zero:

$$F = f = \mu_k mg$$
$$= (0.2)(50)(10) = \underline{100 \text{ N}}$$

Power output: $P = Fv = (100)(1.5) = \underline{\textbf{150 W}}$

Quiz 7.3

1. A 40 kg child slides down a slide, from an initial height of 3 m. The initial speed is 2 m/s. Because of friction, at the bottom of the slide, the child's speed is only 1 m/s. How much heat was generated?

2. A 3 kg block starts from rest and slides down a 30° inclined plane with $\mu_k = 0.2$. It slides a total distance $d = 5$ m. How much work was done by friction?

3. In the previous problem, what is the final speed of the block?

4. An engine propels a rocket through the air at a constant speed of 100 m/s. The engine produces a force of 15 N. What is the power output of the engine?

Chapter summary

Work (J)	$W = F_{//}d$
Kinetic energy (J)	$K = ½mv^2$
Potential energy due to gravity (J)	$U = mgy$
Total energy (J)	$E = K + U$
Energy conservation	Final E = Initial E
If a force such as friction does work W	Final E = Initial $E + W$
Power (J/s, or W)	$P = W/\Delta t = F_{//}v$

End-of-chapter questions

1. A proton of mass 1.67×10^{-27} kg travels at 6×10^4 m/s. An accelerator does 9×10^{-18} J of work on the proton. What is the final speed of the proton?

2. A box slides down an inclined plane. A person tries to prevent it from sliding, by exerting a force of 5 N. However, the box and person slide down the plane a distance of 3 m. How much work did the person do?

3. A block slides down a 30° frictionless inclined plane, starting from rest. It slides a total distance of 4 m. What is its speed at the bottom?

4. After finals, a student opens a window of her dorm room, 10 m above the ground. She throws a water balloon upward, with an initial speed 5 m/s.

    a. What is the speed of the balloon as it falls past the window?
    b. What is the speed of the balloon just before it hits the ground?

5. A 1200 kg car coasts down a 50 m high hill at a constant speed. How much heat was generated by friction?

6. A 5 kg block slides down a 60° inclined plane, starting from rest. It slides a total distance of 4 m. The coefficient of kinetic friction is 0.3.

    a. How much work was done by friction?
    b. What is the speed of the block at the bottom?

7. A boat is powered by an engine that provides a force of 140 N. The boat travels at 10 m/s. What is the power output of the engine?

Challenging problems

1. A wrestler grabs a rhino's horn and pushes up with 14 N of force at a horizontal angle of 44°. The angry rhino pushes the wrestler back 3 m. How much work did the wrestler do?

2. A skier travels down a 35° slope, a total distance of 200 m. The skier's initial speed is 4 m/s. The coefficient of kinetic friction is 0.07.

    a. What is the speed at the bottom of the slope?
    b. At the bottom of the slope, there is a ski jump. The skier is launched into the air. How high does the skier go?

3. Consider a 5 kg block on a 60° inclined plane, with a coefficient of kinetic friction 0.3. A physics professor pushes the block up the plane at a constant speed of 3 m/s. What was the power output of the professor?

4. A 15 kg ball is dropped from a height 7 m. It bounces and reaches a height of only 4 m. How much heat was generated?

8 Momentum

Momentum and impulse

Momentum (**p**) is defined as mass times velocity:

$$\mathbf{p} = m\mathbf{v}$$

Like velocity, **p** is a vector that points in a certain direction.

Force is required to change the momentum of an object. Let $\mathbf{F}_{av}$ be the average force exerted on an object during a time Δt. The change in momentum ($\Delta\mathbf{p}$), also called the *impulse*, is given by

$$\Delta\mathbf{p} = \mathbf{F}_{av}\Delta t$$

1. During a safety test, a 500 kg car travels at 20 m/s in the x direction. It crashes into a wall and comes to rest. What was the change in momentum?

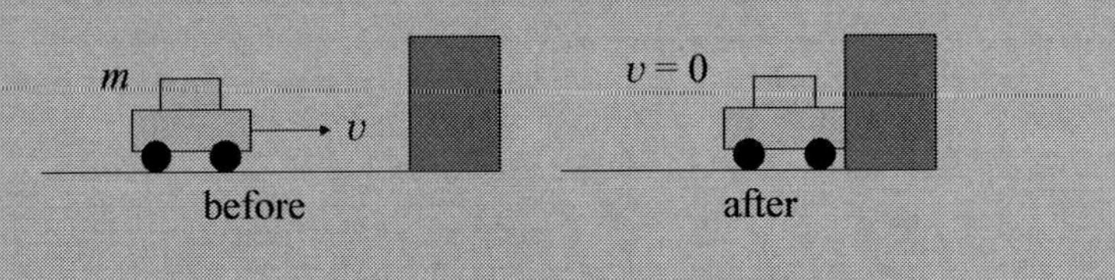

We consider the component of **p** in the x direction.

$p(\text{before}) = mv = (500)(20) = 10{,}000$ kg m/s
$p(\text{after}) = mv = (500)(0) = 0$

$\Delta p = p(\text{after}) - p(\text{before}) = 0 - 10{,}000 =$ **<u>–10,000 kg m/s</u>**

2. In the previous question, the car took 0.2 s to go from 20 m/s to zero. What was the average force exerted by the wall during that time?

$\Delta p = F_{av}\Delta t$
$\Delta p/\Delta t = F_{av}$

$F_{av} = -10{,}000/0.2 =$ **<u>–50,000 N</u>**

During the collision, the wall exerted an average force is 50,000 N in the $-x$ direction.

Inelastic collisions

When objects collide, the momentum of the system is conserved. The sum of the objects' momenta just before the collision equals the sum of momenta just after the collision. *Momentum conservation* is true even when mechanical energy is not conserved. A collision where energy is not conserved is *inelastic*. This happens when objects stick together or there is friction.

Momentum conservation holds as long as there is no *external force* on the system. If two carts collide, but the physics professor holds one to keep it from moving, then we cannot use momentum conservation.

1. A 700 kg car travels at 14 m/s and collides with an identical car that is parked. The cars stick together. What is the speed of the cars immediately after the collision?

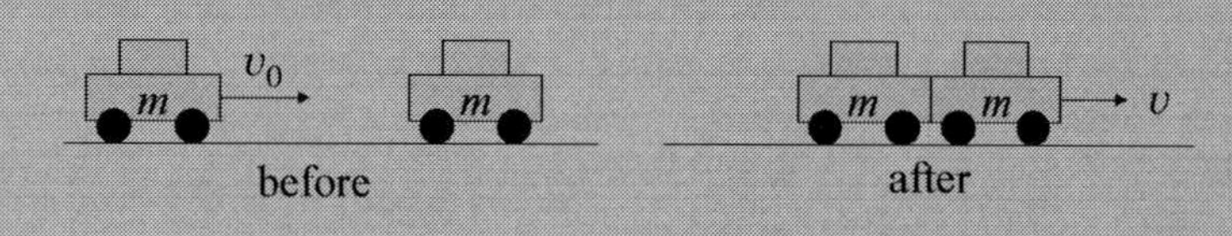

p(before) = mv_0

p(after) = $(2m)v$

p(before) = p(after)

$mv_0 = (2m)v$
$v_0 = 2v$
$v = v_0/2 = 14/2 =$ **7 m/s**

2. In the previous problem, how much mechanical energy was lost?

E(before) = $\frac{1}{2}mv_0^2 = (0.5)(700)(14^2) = 68{,}600$ J

E(after) = $\frac{1}{2}(2m)v^2 = (700)(7^2) = 34{,}300$ J

68,600 – 34,300 = **34,300 J** of energy was lost

The "lost" energy went into heat, sound, and deformation (crumpling) of the cars.

Quiz 8.1

1. A 2000 kg jet travels at 200 m/s in the x direction. An afterburner pushes the jet for 5 s. The jet reaches a speed of 250 m/s. What was the change in momentum?

2. In the previous problem, what was the average force exerted by the afterburner?

3. A person throws a 0.5 kg lump of clay at a speed of 10 m/s. The clay sticks to a 9.5 kg stationary cart. What is the speed of the clay-plus-cart immediately after the collision?

4. In the previous problem, how much kinetic energy was lost?

Elastic collisions

If a collision is *elastic*, then kinetic energy is conserved as well as momentum. Objects such as billiard balls or atoms undergo elastic collisions. They bounce off each other with no loss in energy.

1. A 1 kg mass travels 2 m/s in the x direction. It collides with a 3 kg mass that is initially at rest. After the collision, the 1 kg mass travels 1 m/s in the $-x$ direction. Use momentum conservation to find the velocity of the 3 kg mass.

1 kg 2 m/s → 3 kg — before

1 m/s ← 1 kg 3 kg v → — after

$p = mv$

$p(\text{before}) = (1)(2) + (3)(0) = 2$
$p(\text{after}) = (1)(-1) + 3v = -1 + 3v$

$p(\text{before}) = p(\text{after})$
$2 = -1 + 3v$
$3 = 3v$

$\underline{v = 1 \text{ m/s}}$

2. In the previous problem, calculate the kinetic energy before and after the collision.

$K = ½mv^2$

$K(\text{before}) = ½(1)(2^2) = \underline{\mathbf{2\ J}}$

$K(\text{after}) = ½\,(1)(-1)^2 + ½(3)(1)^2$
$= 0.5 + 1.5 = \underline{\mathbf{2\ J}}$

Since $K(\text{before}) = K(\text{after})$, the collision is elastic.

Collisions in two dimensions

Since momentum is a vector, in an x-y plane, **p** has components in the x and y directions.

$$p_x = mv_x$$
$$p_y = mv_y$$

For a collision in two dimensions, we apply momentum conservation in the x and y directions.

1. A car travels in the x direction at 5 m/s. An identical car travels 10 m/s at an angle 60° to the x axis. The cars collide and stick together. Find their velocity (v_x, v_y) after the collision.

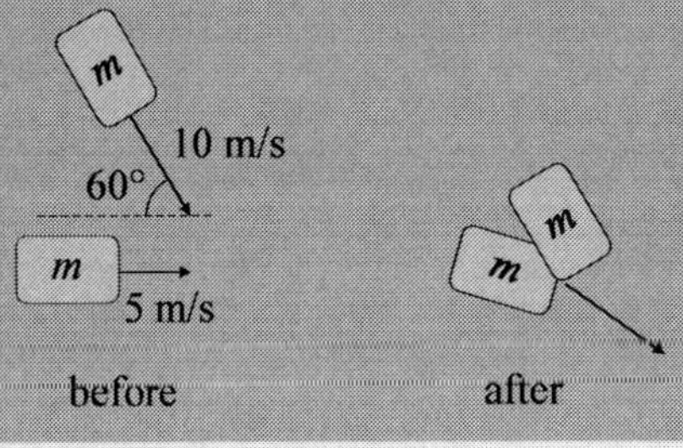

Momentum conservation in the x direction:

$p_x(\text{before}) = m(5) + m(10 \cos 60°) = 5m + 5m = 10m$
$p_x(\text{after}) = (2m)v_x$
$10m = (2m)v_x$
$10 = 2v_x$, or **$v_x = 5$ m/s**

Momentum conservation in the y direction:

$p_y(\text{before}) = m(0) + m(-10 \sin 60°) = -8.66m$
$p_x(\text{after}) = (2m)v_y$
$-8.66m = (2m)v_y$
$-8.66 = 2v_y$, or **$v_y = -4.33$ m/s**

2. In the previous problem, what is the speed and angle of the cars after the collision?

Speed: $v = \sqrt{v_x^2 + v_y^2}$ = **6.6 m/s**

Angle: From trigonometry, $\tan \theta = v_y/v_x$
$\tan \theta = -4.33/5 = -0.866$
$\theta = \tan^{-1}(-0.866) = -41°$

The angle is **41°** below horizontal.

Quiz 8.2

1. A 1 kg mass travels 3 m/s in the x direction. A 5 kg mass travels 3 m/s in the $-x$ direction. They collide. After the collision, the 1 kg mass travels 7 m/s in the $-x$ direction. Use momentum conservation to find the velocity of the 5 kg mass

2. In the previous problem, calculate the kinetic energy before and after the collision.

3. Two identical cars approach an intersection. One car travels 10 m/s in the x direction. The second car travels 6 m/s in the y direction. They collide and stick together. What is their velocity (v_x and v_y) immediately after the collision?

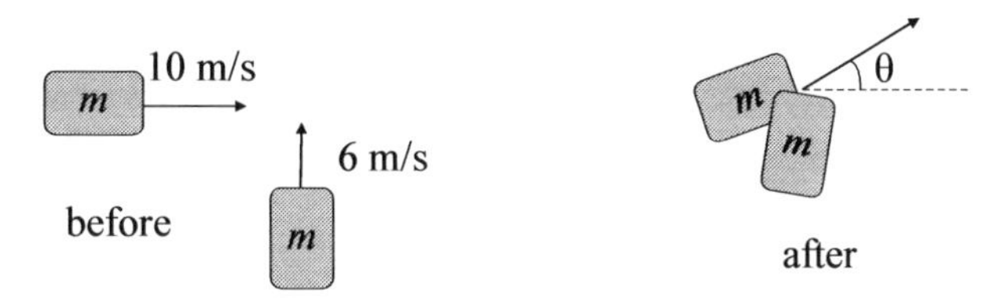

4. In the previous problem, find the speed (v) and angle (θ) after the collision.

The ballistic pendulum

A simple pendulum is a mass that hangs by a "massless" rod or rope. In a ballistic pendulum experiment, a projectile hits the mass and sticks to it. This collision causes the mass swings up to a height h.

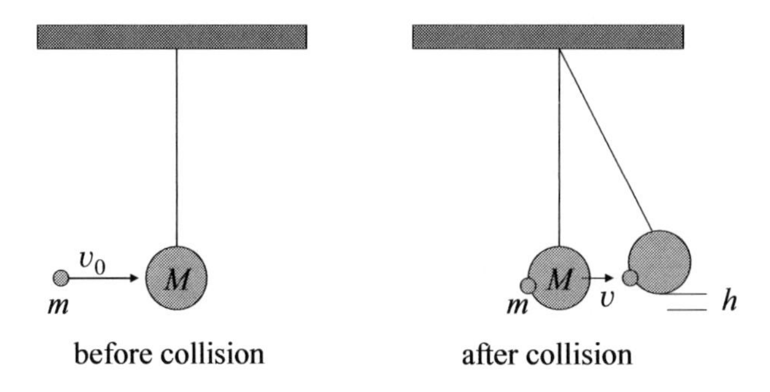

This is a two-part problem. First, calculate the speed of the masses immediately after the collision (v). Second, use energy conservation to calculate the height h.

1. A 0.9 kg ball is suspended from the ceiling by a rope. A 0.1 kg projectile travels at 10 m/s and sticks to the ball. What is the maximum height that the ball will reach?

Before collision: $p = mv_0$
After collision: $p = (m + M)v$

Set them equal: $mv_0 = (m + M)v$

$$(0.1)(10) = (0.1 + 0.9)v$$

$$\underline{1 \text{ m/s} = v}$$

Now, use energy conservation to find h.

Initial $E = ½(m + M)v^2$
Final $E = (m + M)gh$

Set them equal: $½(m + M)v^2 = (m + M)gh$

$$½v^2 = gh$$

$$v^2/2g = h$$

$$h = 1^2/20 = \underline{\mathbf{0.05\ m}}$$

Rockets

A rocket accelerates by expelling gas out a rear nozzle. Because of momentum conservation, the rocket moves in the opposite direction. Consider a rocket of mass M, initially at rest. It expels a mass m in the $-x$ direction at a speed v_{ex}. The rocket moves forward at a speed v.

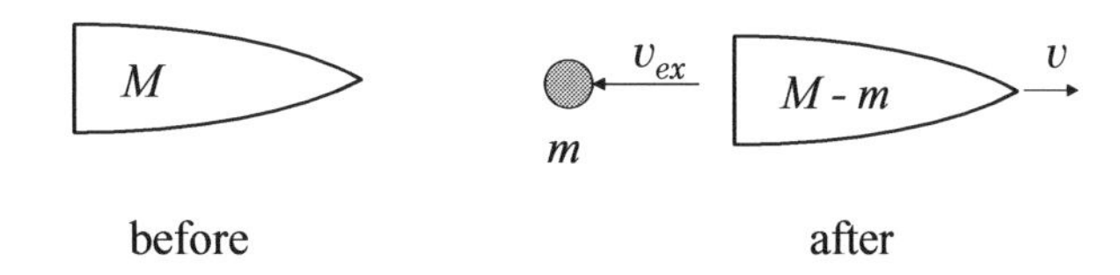

We assume m is much smaller than M. In an actual rocket, the mass of the rocket can decrease significantly as fuel is burned. This gives rise to the "rocket equation," which is beyond the scope of this text.

1. A 1 kg water rocket is launched. It shoots water out the nozzle at a speed 10 m/s. Find the speed of the rocket after it shoots 0.02 kg of water.

$p(\text{before}) = M(0) = 0$
$p(\text{after}) = m(-v_{ex}) + (M - m)v = (0.02)(-10) + (0.98)v$

Set them equal: $0 = (0.02)(-10) + (0.98)v$
$0.2 = (0.98)v$

<u>0.2 m/s</u> $= v$

2. Lester is on a boat floating north at 1 m/s. He shoots a 0.1 kg bullet. The bullet's velocity (relative to the still water) is 500 m/s, south. The mass of Lester, his rifle, and the boat is 250 kg. Find the velocity of the boat after Lester shoots the bullet.

$p(\text{before}) = (250.1)(1)$ (0.1 kg is from the mass of the bullet)
$p(\text{after}) = -(0.1)(500) + 250v$

Set them equal: $250.1 = -50 + 250v$
$300.1 = 250v$

$v =$ **<u>1.2 m/s (north)</u>**

Quiz 8.3

1. A 30 kg block of wood is suspended by a rope. A 0.2 kg projectile travels at 300 m/s and embeds itself in the block. The block swings up to a height h. Find h.

2. A 4500 kg spacecraft is stationary with respect to the International Space Station. It shoots 0.5 kg of gas at a speed 180 m/s. Now what is the speed of the spacecraft?

3. Fred (100 kg) is in a 250 kg canoe that approaches the dock at 0.5 m/s. He jumps toward the dock at a speed 2 m/s (relative to the dock). Now what is the velocity of the canoe?

Chapter summary

Momentum	$\mathbf{p} = m\mathbf{v}$
Impulse	$\Delta\mathbf{p} = \mathbf{F}_{av}\Delta t$
Momentum conservation	$p(\text{before}) = p(\text{after})$
In two dimensions	$p_x = mv_x$ $p_y = mv_y$

End-of-chapter questions

1. A 0.5 kg basketball hits the floor with a speed 7 m/s. It is in contact with the floor for 0.02 s. It bounces back in the opposite direction, with a speed 5 m/s. What was the average force exerted by the floor?

2. A 2 kg cart and a 1 kg cart approach each other, each with a speed 3 m/s. They collide and stick together.

   a. Find the speed of the carts after the collision.
   b. How much mechanical energy was lost?

3. A car rolls down a 20 m high frictionless hill, starting from rest. At the bottom, it collides with an identical parked car. They stick together. What is their speed?

4. A helium nucleus (6.64×10^{-27} kg) collides elastically with a stationary silicon nucleus (4.65×10^{-26} kg). The initial speed of the helium nucleus is 4.62×10^4 m/s. It scatters at 90° with a speed 4.00×10^4 m/s. Find the speed (v) and angle (θ) of the silicon nucleus after the collision.

5. A 45 kg ball is suspended by a rope. A 30 kg child runs 2 m/s and grabs on to the ball. How high will the child and ball swing?

6. The Death Star has a mass of a small moon, 10^{20} kg. Assume it is at rest. It expels garbage of mass 10^9 kg, with a speed 2000 m/s. What is the speed of the Death Star after expelling the garbage?

Challenging problems

1. A cue ball travels 10 m/s in the x direction. It collides with an "8 ball" and travels with a speed v_2. The 8 ball travels with a speed v_1. Find v_1 and v_2.

2. A 0.1 kg bullet travels at 300 m/s and passes through a 40 kg block of wood that is suspended by a rope. The bullet emerges from the block with a speed 100 m/s. The block swings to a maximum height h. Find h.

3. Lester is in a stationary boat with his 50 kg dog. The mass of Lester, his rifle, and the boat is 250 kg. Lester shoots a 0.1 kg bullet with a velocity of 500 m/s, south. The spooked dog jumps off the boat at a velocity 5 m/s, northwest. Find the speed of the boat.

4. A 1 kg object moves at 9 m/s toward a 3 kg object. They collide and are motionless.

    a. What was the velocity of the 3 kg object before the collision?
    b. How much mechanical energy was lost during the collision?

9 Rotation

Angular velocity

An object rotates about an *axis*. Some objects, like the earth, rotate slowly. Others, like a spinning top, rotate quickly. *Angular velocity* (ω) is measured in radians per second (rad/s), where a complete rotation is 2π radians. For a rigid object, all parts of the object have the same ω.

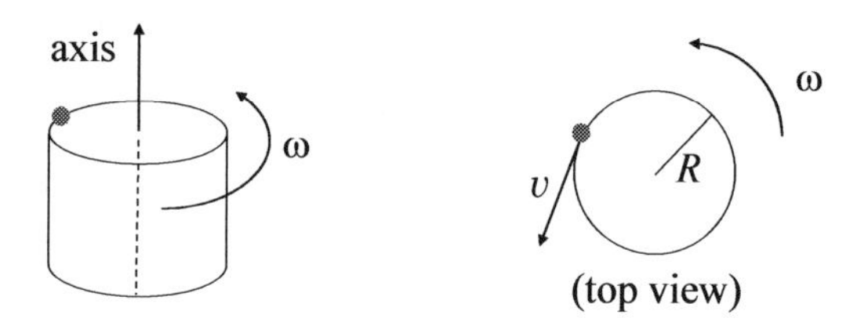

A point on the object has a tangential velocity v (Chapter 6). It is related to ω by

$$v = \omega R$$

where R is the distance from the point to the axis.

1. A wheel of 0.5 m radius rotates about a fixed axis at 120 rotations per minute (rpm). What is the angular velocity?

Convert rpm to rad/s:

$$\omega = \frac{120 \text{ rotations}}{\text{min}} \frac{1 \text{ min}}{60 \text{ s}} \frac{2\pi \text{ rad}}{\text{rotation}} = \underline{\mathbf{4\pi \text{ rad/s}}}$$

2. In the previous problem, a piece of gum is stuck to the wheel. What is its tangential velocity?

$$v = \omega R = (4\pi)(0.5) = \underline{\mathbf{6.28 \text{ m/s}}}$$

Rotational kinetic energy

The kinetic energy due to rotation is given by

$$K = \tfrac{1}{2}I\omega^2$$

where I is the *moment of inertia*, which has units kg·m². Its value depends on the shape and density distribution of the object. A "uniform" object has a constant density throughout.

In the next section, we will see how to calculate I.

1. A uniform disk has a moment of inertia given by $I = 1/2\ MR^2$. Consider a plastic disk with $M = 2$ kg and $R = 0.4$ m. It spins at 19.1 rpm about its fixed central axis. Find the kinetic energy.

$I = 1/2\ MR^2 = (0.5)(2)(0.4^2) = \underline{0.16\ \text{kg·m}^2}$

$$\omega = \frac{19.1\ \text{rotations}}{\text{min}}\ \frac{1\ \text{min}}{60\ \text{s}}\ \frac{2\pi\ \text{rad}}{\text{rotation}} = \underline{2\ \text{rad/s}}$$

$K = \tfrac{1}{2}I\omega^2 = (0.5)(0.16)(2^2) = \underline{\mathbf{0.32\ J}}$

2. A uniform sphere has a moment of inertia given by $I = 2/5\ MR^2$. A neutron star can be modeled as an extremely dense uniform sphere. Consider a neutron star of mass 10^{30} kg and radius 10 km. It spins at 5 rotations per second. What is its kinetic energy?

$I = 2/5\ MR^2 = (0.4)(10^{30})(10{,}000^2) = \underline{4 \times 10^{37}\ \text{kg·m}^2}$

$$\omega = \frac{5\ \text{rotations}}{\text{s}}\ \frac{2\pi\ \text{rad}}{\text{rotation}} = \underline{31.4\ \text{rad/s}}$$

$K = \tfrac{1}{2}I\omega^2 = (0.5)(4 \times 10^{37})(31.4^2) = \underline{\mathbf{2 \times 10^{40}\ J}}$

Quiz 9.1

1. The earth rotates once every 24 hours. What is its angular velocity?

2. The radius of the earth is approximately 6400 km. Find the tangential velocity of a person on the equator. (Ignore the motion of the earth around the sun).

3. The earth can be modeled as a uniform sphere ($I = 2/5\ MR^2$) of mass 6×10^{24} kg. Find its rotational kinetic energy.

4. Consider an optical disc ($I = 1/2\ MR^2$) of mass 16 g and diameter 120 mm. It spins at 240 rpm. Find its kinetic energy.

Moment of inertia

For an object like a disk or sphere, we divide the object into little masses: m_1, m_2, etc. Each mass is some perpendicular distance (r_1, r_2, etc.) from the axis.

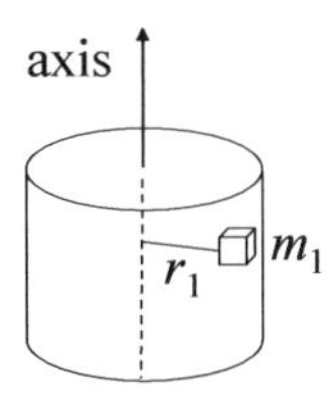

The moment of inertia is defined as

$$I = m_1r_1^2 + m_2r_2^2 + \ldots$$

1. Find the moment of inertia for a hoop of mass M and radius R.

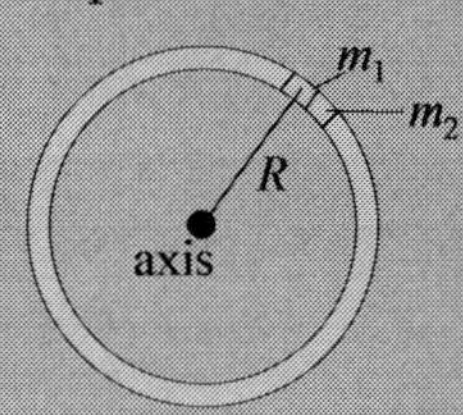

$$\begin{aligned} I &= m_1r_1^2 + m_2r_2^2 + \ldots \\ &= m_1R^2 + m_2R^2 + \ldots \\ &= (m_1 + m_2 + \ldots)R^2 \end{aligned}$$

(All the little masses have the same R)

$\underline{\boldsymbol{I = MR^2}}$ (The sum of the little masses is M)

2. Two small masses (0.1 kg) are connected by a massless rod of length 0.5 m. Calculate the moment of inertia if the rotational axis is in the center of the rod.

0.5 m

$I = (0.1)(0.25^2) + (0.1)(0.25^2) =$ **<u>0.0125 kg·m²</u>**

3. In the previous problem, calculate I if the axis goes through one of the masses.

$I = (0.1)(0^2) + (0.1)(0.5^2) =$ $\underline{\mathbf{0.025\ kg{\cdot}m^2}}$

Torque

Torque is a kind of twisting force that can cause objects to rotate. When you open a door, you exert a torque on the door that makes it rotate about its hinges. The rotational axis is called the *pivot* (for a door, it is the hinges). The magnitude of torque (τ) is

$$\tau = rF \sin \theta$$

where r is the distance from the pivot to the point where the force F is applied, and θ is the angle between **r** and **F**. Torque has units N·m.

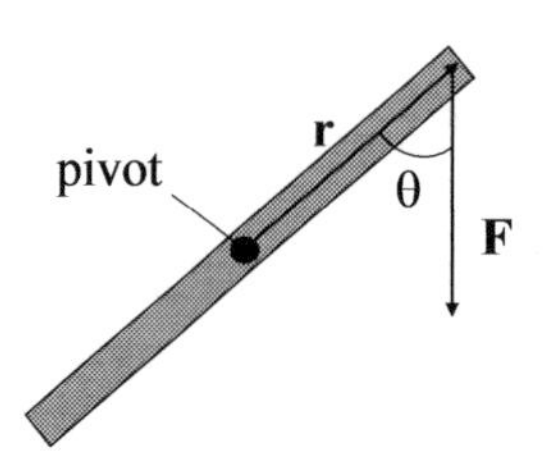

By convention, τ is positive (+) if it pushes the object in a counterclockwise direction, and negative (–) if it pushes in a clockwise direction.

1. Find the torque when a person exerts 10 N on the edge of a door.

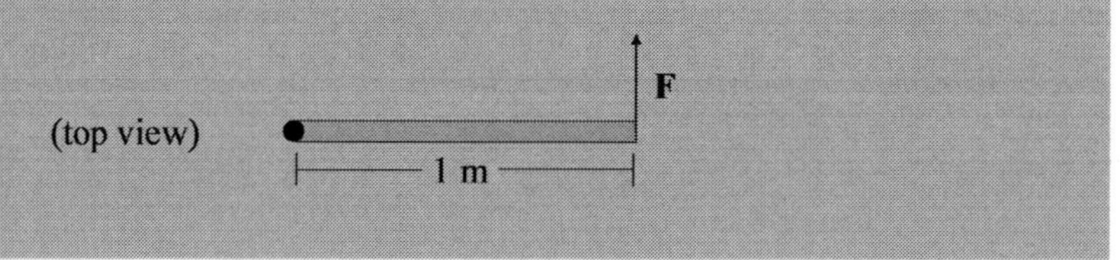

$\tau = rF \sin \theta = rF \sin 90° = rF$

$\tau = (1)(10) =$ **10 N·m** (+ because it makes the door go counterclockwise)

2. Find the torque when a person exerts 10 N on the center of the door.

F

0.5 m

$\tau = rF = (0.5)(10) =$ **5 N·m**

Quiz 9.2

1. A hoop has a diameter of 1 m and mass 0.6 kg. It rotates about its center. What is the moment of inertia?

2. Two small masses, 1 kg and 3 kg, are connected by a massless rod of length 4 m. The rotational axis is 1 m from the heavier mass. Find the moment of inertia.

3. A person pushes the edge of a door, with a force of 8 N, at a 30° angle. What is the torque?

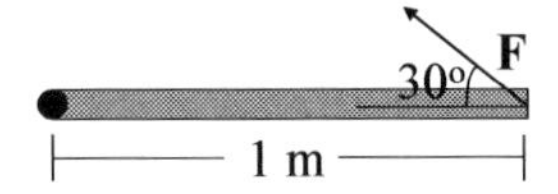

4. A driver wants to make a left turn. The steering wheel has a 0.2 m radius. Her left hand exerts a 3 N force down while her right hand exerts a 3 N force up. Find the torque.

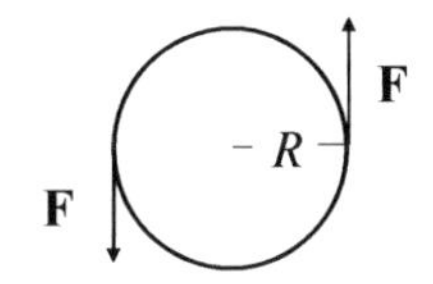

Turning a wrench

Torque is important if you are using a wrench to loosen or tighten a bolt. In these problems, the bolt is the pivot.

1. A worker exerts 7 N of downward force on a wrench. The wrench is 0.2 m long and is oriented at 25° to horizontal. Calculate the torque.

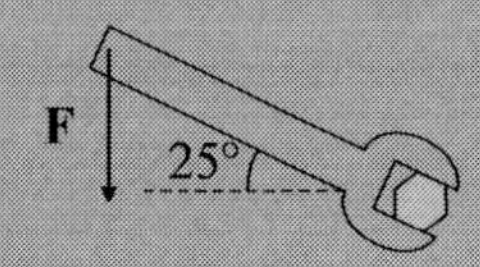

$r = 0.2$ m
$F = 7$ N
$\theta = 90° - 25° = 65°$ (θ is the angle between **r** and **F**)

$\tau = rF \sin\theta$
$= (0.2)(7)\sin(65°) = \underline{\mathbf{1.3\ N{\cdot}m}}$

The sign is *positive* because the torque makes the wrench rotate counterclockwise.

2. A worker exerts 8 N of force at a horizontal angle of 30°. The wrench is vertical and has a length of 0.1 m. Find the torque.

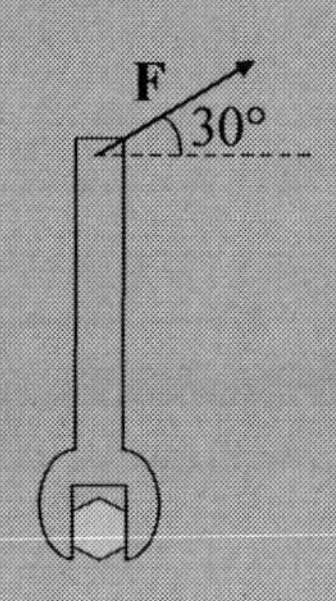

$\theta = 30° + 90° = 120°$

$\tau = rF \sin\theta$
$= (0.1)(8)\sin(120°) = 0.7\ \text{N}\cdot\text{m}$

The sign is *negative* because the torque makes the wrench rotate clockwise.

So, $\tau = \underline{\mathbf{-0.7\ N{\cdot}m}}$

Angular acceleration

Recall that acceleration (a) is change in velocity versus time and is measured in m/s^2 (Chapter 1). *Angular acceleration* (α) is change in angular velocity versus time and is measured in rad/s^2:

$$\alpha = \Delta\omega / \Delta t$$

Torque causes angular acceleration:

$$\tau = I\alpha$$

This is the rotational analog of $F = ma$.

1. A merry-go-round (I = 20 kg·m^2) spins at 0.1 rad/s. A torque of 10 N·m is applied, making it spin faster. How long must the torque be applied, to make it spin at 2.1 rad/s?

$\tau = I\alpha$
$10 = (20)\alpha$, or $\underline{\alpha = 0.5 \text{ rad/s}^2}$

$\alpha = \Delta\omega/\Delta t$
$0.5 = 2/\Delta t$ ($\Delta\omega = 2.1 - 0.1 = 2$ rad/s)
$\Delta t = 2/0.5 = \underline{\mathbf{4\ s}}$

2. String is wrapped around the inner diameter (r = 0.03 m) of a spool. The spool has a moment of inertia $I = 1/2\ MR^2$, where M = 0.5 kg and R = 0.1 m. It is free to rotate about a fixed axis. A cat pulls the string with 2 N of force. Find the angular acceleration.

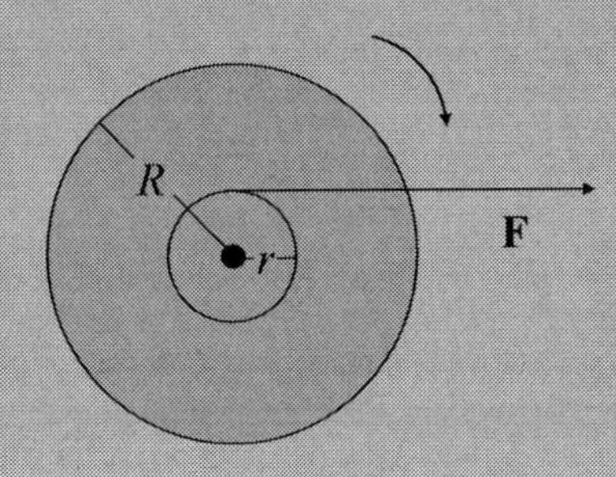

$I = 1/2\ MR^2 = (0.5)(0.5)(0.1^2) = \underline{0.0025 \text{ kg·m}^2}$

$\tau = rF = -(0.03)(2) = -0.06$ N·m (– because it pulls the spool in a clockwise direction)

$\tau = I\alpha$
$-0.06 = (0.0025)\alpha$, or $\alpha = \underline{\mathbf{-24\ rad/s^2}}$ (– means it spins faster in a clockwise direction)

Quiz 9.3

1. A worker exerts a downward force of 10 N on a wrench that is 30° to horizontal. The length of the wrench is 0.15 m. Find the torque.

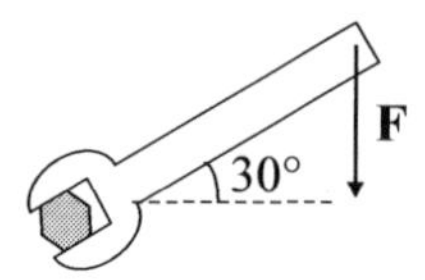

2. A steel rod hangs vertically. A force of 50 N is applied, 35° to horizontal, 1.5 m from the pivot. Calculate the torque.

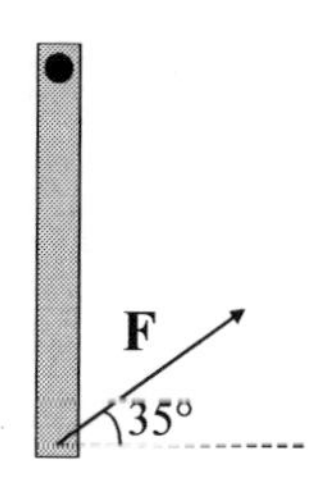

3. Bobby has a globe ($I = 2/5\ MR^2$, $M = 1$ kg, $R = 0.2$ m) that is spinning at 4 rad/s. He puts his finger on the globe, and it stops spinning in 3 s. What was the magnitude of the torque that Bobby applied?

4. Rope is wound around a cylinder ($I = 1/2\ MR^2$, $M = 48$ kg, $R = 0.5$ m) that is free to rotate about a fixed axis. A sailor pulls the rope with a force of 6 N. Find the angular acceleration.

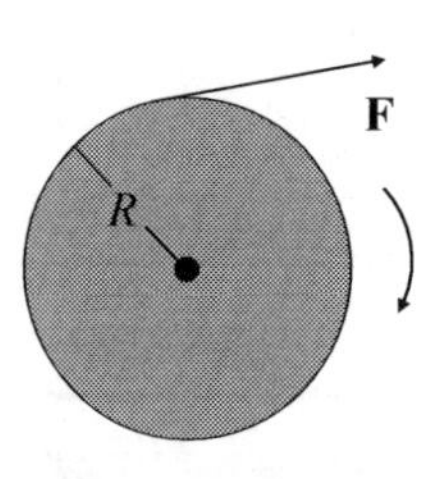

Chapter summary

Angular velocity (ω, rad/s) Tangential velocity (v, m/s)	$v = \omega R$
Rotational kinetic energy (J)	$K = \frac{1}{2} I\omega^2$
Moment of inertia (kg·m^2)	$I = m_1 r_1^2 + m_2 r_2^2 + \ldots$
Torque (N·m)	$\tau = rF \sin\theta$
Angular acceleration (α, rad/s^2)	$\alpha = \Delta\omega/\Delta t$ $\tau = I\alpha$
Counterclockwise = positive, clockwise = negative.	

End-of-chapter questions

1. A merry-go-round of diameter 4 m rotates at 15 rpm.

   a. What is its angular velocity?
   b. A child is on the edge. What is the child's velocity?

2. An airplane is preparing for takeoff. Its propeller ($I = 1/3\ MR^2$) rotates at 1200 rpm. The mass of the propeller is 100 kg and its radius is 1 m. Find the rotational kinetic energy.

3. A 3 kg and 5 kg mass are attached by a 4 m massless rod. They rotate at 4 rad/s about a fixed central axis. Calculate the kinetic energy.

4. At a factory, a very long (3 m) wrench is used to tighten a large bolt. The wrench is at 40° to horizontal. A worker (100 kg) tries to turn the wrench by hanging from the end. What is the torque?

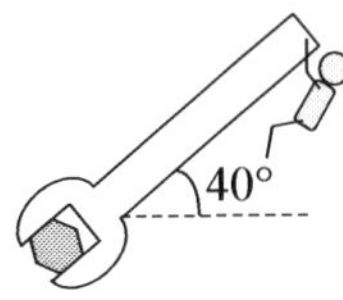

5. Two 2 kg masses are connected by a massless rod of length 1.5 m. At the rod's center is massless cylinder ($r = 0.4$ m) with a central axis. String is wound around the cylinder. A person pulls on the string with a force of 5 N. Find the angular acceleration.

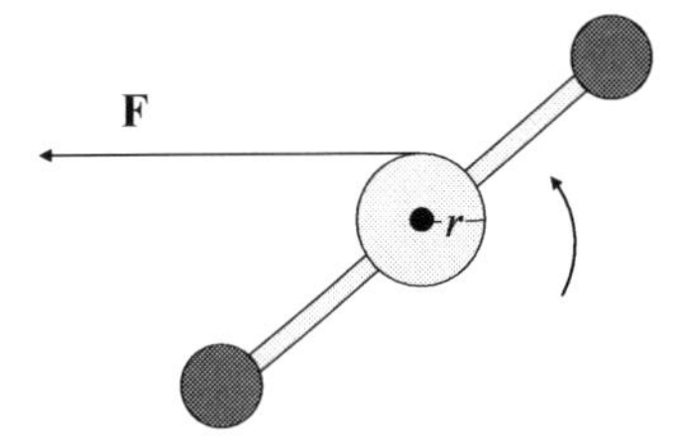

Challenging problems

1. A massless rod of length $2a$ ($a = 2$ m) has identical masses ($m = 3$ kg) attached. The masses are a and $2a$ from the rotation axis. The rod is tilted 60° above horizontal.

   a. What is the torque about the rotation axis?
   b. After some time, the rod and masses rotate at 6 rpm. What is the kinetic energy?

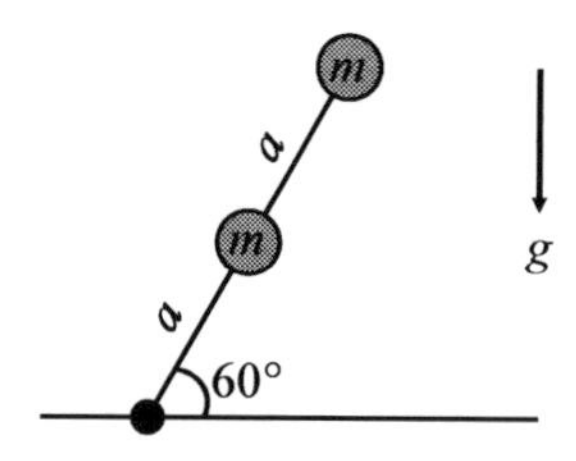

2. A ball of mass 6 kg and radius 0.3 m spins at 8 rad/s. A person catches it. The sliding friction from the person's hands makes the ball stop spinning in 0.1 s. What was the average torque exerted by the hands? Model the ball as a uniform sphere, $I = 2/5\ MR^2$.

3. Rope is wound around a cylinder ($I = 1/2\ MR^2$) of mass 64 kg and radius 0.5 m. The cylinder is initially at rest. A sailor pulls the rope with a force of 8 N over a distance 2 m.

   a. How much work was done by the sailor?
   b. What is the angular velocity of the cylinder?

4. Two identical twins (30 kg each) are on a merry-go-round. One twin is 1 m from the center and the second twin is 2 m from the center. The merry-go-round is a disk of radius 3 m and has a mass 100 kg, $I = 1/2\ MR^2$. The merry-go-round spins at 15 rpm.

   a. Find the moment of inertia for the merry-go-round with the children on it.
   b. Calculate the kinetic energy.

10 Angular momentum

Angular momentum due to rotation

Suppose we have a disk rotating about a fixed central axis. Because the axis is fixed, the object has zero momentum **p** (Chapter 8). However, it does have *angular momentum* (L), defined as

$$L = I\omega$$

This is the rotational analog of $\mathbf{p} = m\mathbf{v}$.

By convention, L is positive (+) if the rotation is counterclockwise, and negative (–) if it is clockwise.

1. A uniform disk ($I = 1/2\ MR^2$) of mass 0.1 kg and radius 0.1 m spins at 120 rpm in a clockwise direction. Find the angular momentum.

$I = 1/2\ MR^2 = (0.5)(0.1)(0.1^2) = \underline{5 \times 10^{-4}\ \text{kg}\cdot\text{m}^2}$

$$\omega = \frac{120\ \text{rotations}}{\text{min}}\frac{1\ \text{min}}{60\ \text{s}}\frac{2\pi\ \text{rad}}{\text{rotation}} = \underline{12.6\ \text{rad/s}}$$

$L = I\omega = -(5 \times 10^{-4})(12.6) = \underline{\mathbf{-0.0063\ kg\cdot m^2/s}}$ (– because rotation is clockwise)

2. A small mass m travels in a circle of radius R with a tangential velocity v. Calculate the angular momentum.

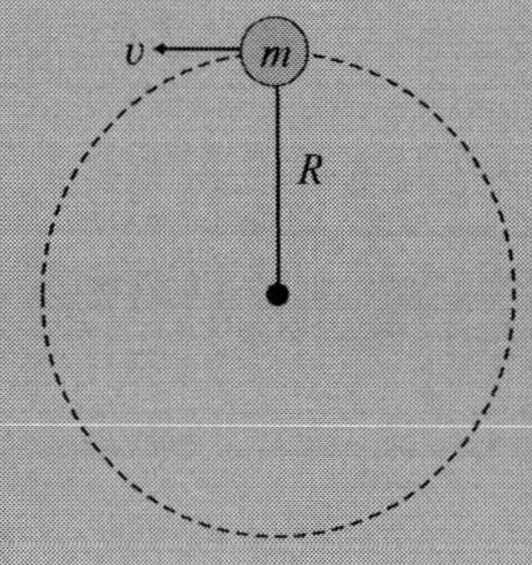

The moment of inertia is $\underline{I = mR^2}$.

The tangential velocity is $v = \omega R$, or $\underline{\omega = v/R}$.

$L = I\omega = (mR^2)(v/R) = \underline{\mathbf{mvR}}$

This is true for any small mass going in a circle of radius R.

Conservation of angular momentum

If there is no external torque on a rotating object, then angular momentum is conserved. For example, an ice skater brings in her arms, reducing her moment of inertia (I). In order to conserve angular momentum ($L = I\omega$), her angular velocity (ω) must increase.

1. A massless rod of length 1.0 m has two masses attached to it. It rotates at 3 rad/s. The rod expands to a length of 2.0 m. Find the angular velocity.

$I = mR^2 + mR^2 = \underline{2mR^2}$

Before: $L = I\omega = 2m(0.5^2)(3) = 1.5m$

After: $L = I\omega = 2m(1.0^2)\omega = 2m\omega$

Set them equal: $1.5m = 2m\omega$

$1.5 = 2\omega$

$\underline{\mathbf{0.75\ rad/s}} = \omega$

2. A star ($I = 2/5\ MR^2$) of radius 10^6 km rotates at 1×10^{-5} rad/s. It collapses to a neutron star, which has a radius of 10 km. Find the angular velocity assuming no loss of mass.

Before: $L = 2/5\ M(10^9\ \text{m})^2(1 \times 10^{-5}\ \text{rad/s}) = 4 \times 10^{12}\ M$

After: $L = 2/5\ M(10^4\ \text{m})^2\omega = 4 \times 10^7\ M\omega$

Set them equal: $4 \times 10^{12}\ M = 4 \times 10^7\ M\omega$

$4 \times 10^{12} = 4 \times 10^7\ \omega$

$\underline{\mathbf{1 \times 10^5\ rad/s}} = \omega$

Quiz 10.1

1. Consider Earth to be a uniform sphere ($I = 2/5\ MR^2$) of mass 6×10^{24} kg and radius 6400 km. It rotates once every 24 hours. What is the angular momentum (as viewed from a point above the North Pole)?

2. A race car of mass 1000 kg travels at 30 m/s around a circular track of radius 250 m. The car travels in a clockwise direction. Find the angular momentum.

3. A person of negligible mass holds two 50 kg masses at arm's length (0.8 m from the center). The person sits on a massless chair that rotates at 1.5 rad/s. The person brings his arms in so the masses are only 0.2 m from the center. What is the angular velocity?

4. In the previous problem, suppose the stool has a moment of inertia 36 kg·m^2. (This must be added to the moment of inertia of the two 50 kg masses.) Find the angular velocity.

Spinning disk

These problems involve the angular momentum conservation of a spinning disk plus a small mass. The important point is to calculate the angular momentum due to the disk *and* the mass.

1. A uniform disk ($I = 1/2\ MR^2$) of mass 4 kg and radius 0.5 m spins at $\omega_0 = 3$ rad/s. A piece of clay ($m = 0.25$ kg) is dropped onto the edge of the disk. Find the angular velocity of the disk + clay.

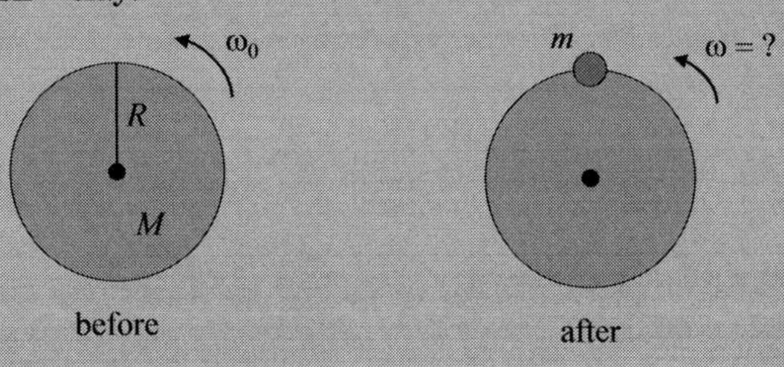

Before: $I = 1/2\ MR^2 = (0.5)(4)(0.5^2) = 0.5$
$L = I\omega_0 = (0.5)(3) = \underline{1.5}$

After: $I = 1/2\ MR^2 + mR^2 = (0.5)(4)(0.5^2) + (0.25)(0.5^2) = 0.5625$
$L = I\omega = \underline{0.5625\ \omega}$

Set them equal: $1.5 = 0.5625\ \omega$
$\underline{\mathbf{2.7\ rad/s}} = \omega$

2. A person ($m = 50$ kg) is on a merry-go-round ($I = 1/2\ MR^2$, $M = 100$ kg, $R = 3$ m), initially at rest. The person runs at a tangential velocity 3 m/s, relative to the ground. Find the angular velocity of the merry-go-round.

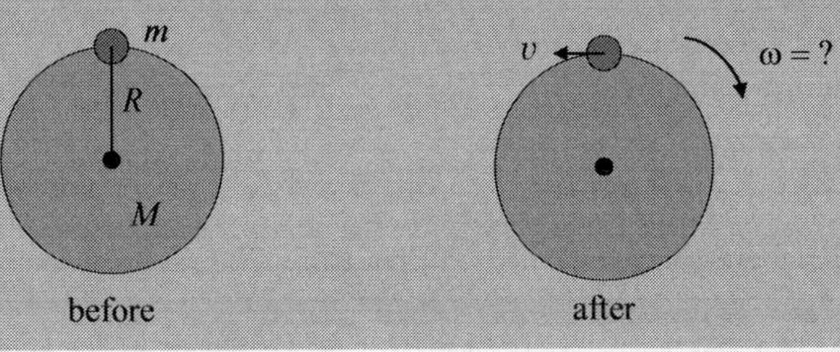

Before: $\underline{L = 0}$ (nothing is moving)

After: $L(\text{person}) = mvR = (50)(3)(3) = 450$
$L(\text{merry-go-round}) = -1/2\ MR^2\omega = -(0.5)(100)(3^3)\omega = -450\ \omega$
Total $\underline{L = 450 - 450\omega}$

Set them equal: $0 = 450 - 450\omega$
$\omega = \underline{\mathbf{1\ rad/s}}$

Travelling object

Suppose a ball is on a string (ignore gravity). It travels in a circle of radius R with a tangential velocity v, in a clockwise direction. The angular momentum is $L = -mvR$. Suddenly, the string breaks. The ball continues traveling at a velocity v. The angular momentum is unchanged. It is still $L = -mvR$.

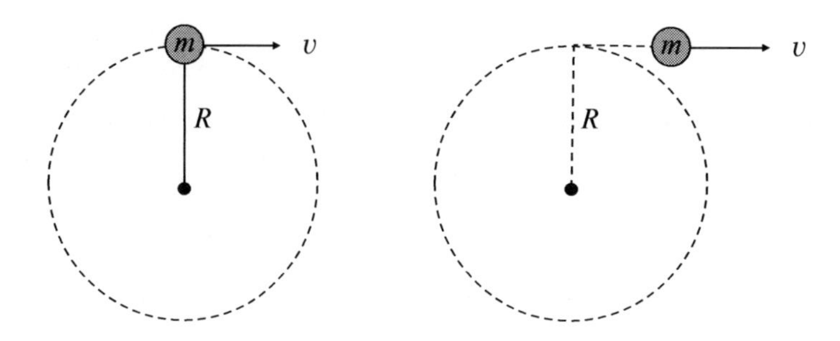

The magnitude of angular momentum for an object travelling at a speed v is

$$L = mvR$$

where R is the perpendicular distance to the axis. (If you extrapolate the object's path, R is the closest distance to the axis.) L is positive (+) if the motion has a counterclockwise orientation, and negative (–) if the motion is counterclockwise.

1. A 0.1 kg piece of clay travels at a speed 5 m/s. It hits a propeller (I = 1/3 MR^2, M = 3 kg, R = 0.2 m) and sticks to it. The propeller rotates on a fixed axis. Find the angular velocity.

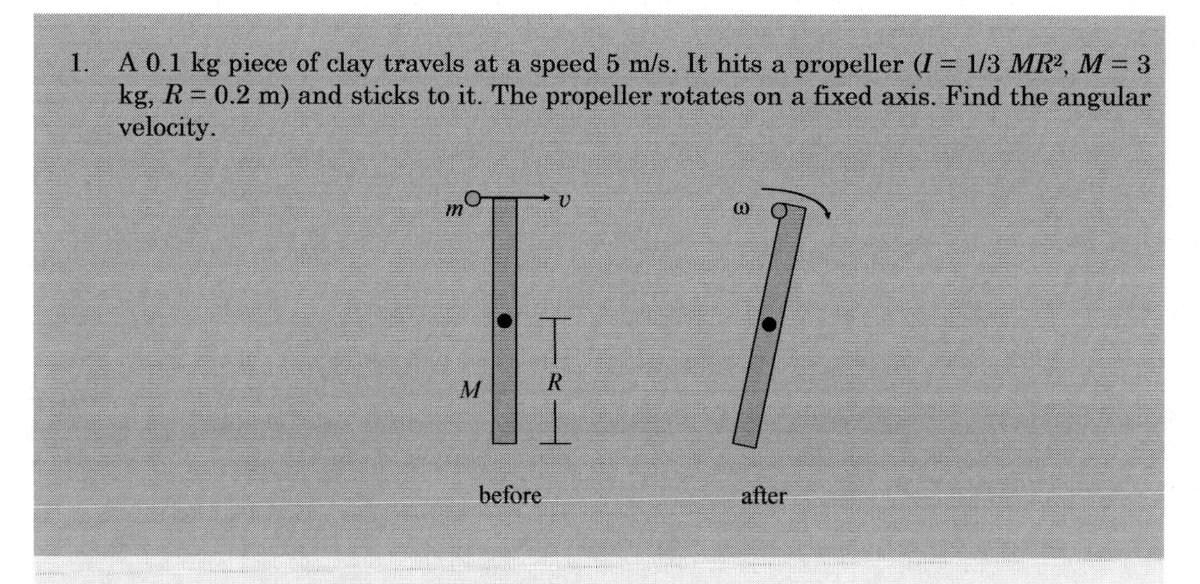

Before: $L = -mvR = -(0.1)(5)(0.2) = \underline{-0.1}$

After: $I = 1/3\ MR^2 + mR^2 = (1/3)(3)(0.2^2) + (0.1)(0.2^2) = 0.044$
$L = -I\omega = \underline{-0.044\omega}$

Set them equal: $-0.1 = -0.044\omega$
<u>2.3 rad/s</u> = ω

Quiz 10.2

1. A 75 kg person is at the center of a 150 kg merry-go-round ($I = 1/2\ MR^2$, $R = 2$ m), which spins at 1.5 rad/s. The person walks to the edge of the merry-go-round. Now what is the angular velocity?

2. A 40 kg child is on a 150 kg merry-go-round ($I = 1/2\ MR^2$, $R = 2$ m), initially at rest. She runs in a circle of radius 1 m, with a speed 0.5 m/s relative to the ground. Calculate the angular velocity of the merry-go-round.

3. A 100 kg person jumps, with a tangential velocity of 3 m/s, onto the edge of a stationary 200 kg merry-go-round ($I = 1/2\ MR^2$, $R = 2$ m). Find the angular velocity after the person lands on the merry-go-round.

Object on a string

Here is an example of angular momentum conservation, where we consider how force is required to pull an object into a tighter orbit.

1. An object of mass 0.1 kg is on a frictionless horizontal table. It travels in a circle of radius 0.1 m with a tangential velocity 2 m/s. It is attached to a string that goes through a hole in the table. A force F is exerted on the string. Find F.

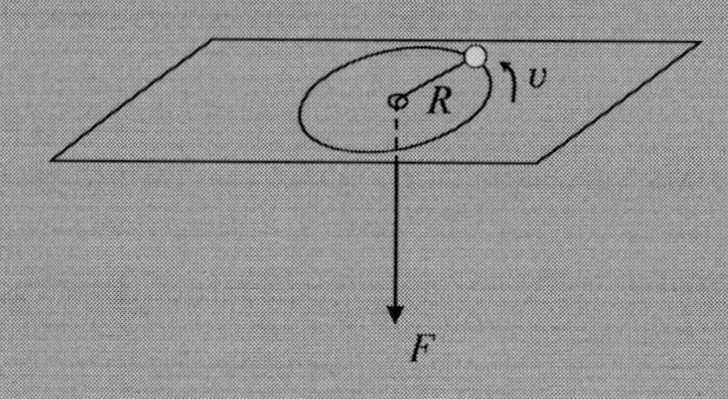

$F = ma$
$a = v^2/R$ (Chapter 6)

$F = mv^2/R$
$= (0.1)(2^2)/0.1 = \underline{\mathbf{4\ N}}$

2. In the previous problem, suppose the force is increased to 32 N. Find the radius R.

$F = mv^2/R$
$32 = (0.1)v^2/R$
$\underline{320 = v^2/R}$

Now, use angular momentum conservation, where $L = mvR$.

$L(\text{before}) = (0.1)(2)(0.1) = 0.02$
$L(\text{after}) = (0.1)vR$
$0.02 = (0.1)vR$
$0.2 = vR$, or $\underline{v = 0.2/R}$

Plug this into the first underlined equation: $320 = (0.2/R)^2/R$
$320 = 0.04/R^3$
$8000 = 1/R^3$
$20 = 1/R$
$\underline{\mathbf{0.05\ m}} = R$

Two rotating objects

These angular momentum conservation problems involve two rotating objects.

1. A physics student sits on a stool that is free to rotate. The moment of inertia of the student and stool is 80 kg·m^2. The student holds a bike wheel (I = 1 kg·m^2) that rotates at 20 rad/s. From the top view, it rotates counterclockwise. The student flips the wheel so it rotates clockwise at 20 rad/s. Find the angular velocity of the student.

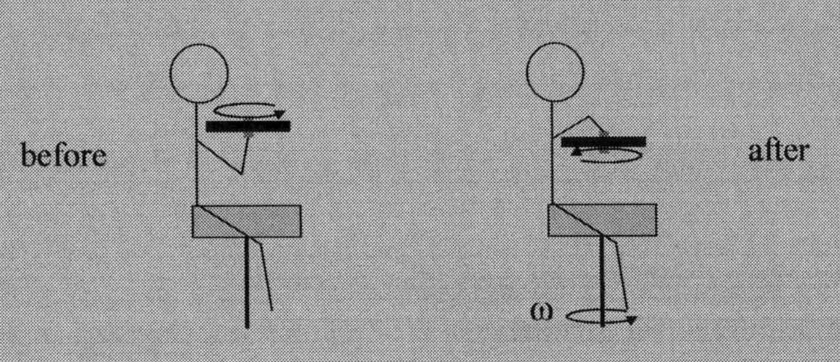

$L = I\omega$
L(before) = (1)(20) = 20
L(after) = –(1)(20) + (80)ω = –20 + 80ω

Set them equal: 20 = –20 + 80ω
40 = 80ω
0.5 rad/s = ω

2. Two uniform disks ($I = 1/2\ MR^2$) are spinning about the same central axis. They both have a radius of 0.1 m. The first disk is 2 kg and spins clockwise at 15 rpm. The second disk is 4 kg and spins counterclockwise at 40 rpm. They collide and stick together. What is their rotational speed (in rpm)?

First disk: $I = (0.5)(2)(0.1^2) = 0.01$
Second disk: $I = (0.5)(4)(0.1^2) = 0.02$

$L = I\omega$
For simplicity, we will keep ω in units of rpm.

L(before) = –(0.01)(15) + (0.02)(40) = 0.65
L(after) = (0.01 + 0.02)ω = (0.03)ω

Set them equal: 0.65 = (0.03)ω
22 rpm = ω

Quiz 10.3

1. A 0.1 kg puck travels in a circle of radius 0.10 m on a frictionless table. It is attached to a string that goes down a hole and is attached to two 0.2 kg masses. What is the speed of the puck?

2. In the previous problem, one of the masses is removed. Now what is the radius of the circle?

3. A toy helicopter has a moment of inertia 0.1 kg·m^2 (not including the rotor). Its rotor (I = 0.004 kg·m^2) spins at 50 rad/s. The rotor slows to 10 rad/s. This causes the helicopter to rotate. Find the angular velocity of the helicopter.

4. A uniform disk ($I = 1/2\ MR^2$, $M = 2$ kg, $R = 0.5$ m) rotates at 120 rpm. A second uniform disk ($M = 4$ kg, $R = 0.5$ m) is initially not rotating. The second disk lands on top of the first disk. Find the rotational speed (rpm) of the two disks.

Chapter summary

Angular momentum due to rotation	$L = I\omega$
Angular momentum of a travelling object	$L = mvR$
Angular momentum conservation	L(before) = L(after)
Counterclockwise = positive, clockwise = negative.	

End-of-chapter questions

1. A globe ($I = 2/5\ MR^2$, $M = 0.25$ kg, $R = 0.316$ m) is in the classroom. A student puts a 0.1 kg piece of clay on the equator. The student spins the globe counterclockwise at 19.1 rpm. Calculate the angular momentum.

2. Two small masses are in a 1 m long tube of negligible mass. The masses are connected to the central axis by two 0.2 m long strings. The tube rotates at 8 rad/s. One string breaks, causing the mass to slide to the end of the tube. Calculate the angular velocity.

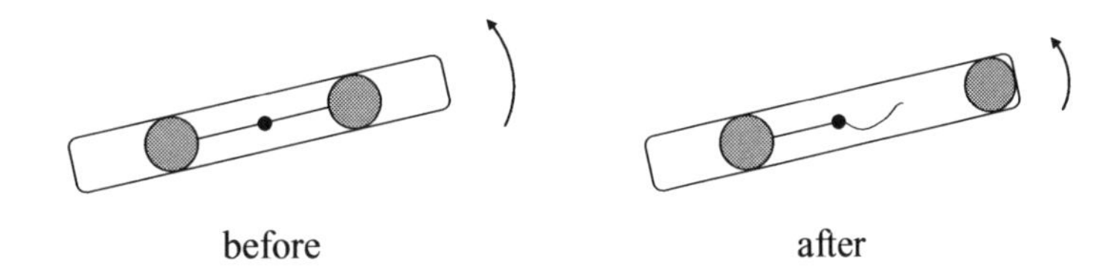

3. A 25 kg child is on the edge of a merry-go-round ($I = 1/2\ MR^2$, $M = 100$ kg, $R = 3$ m) that spins at 20 rpm. She walks to the center. Now what is the rotational speed of the merry-go-round (rpm)?

4. A disk ($I = 1/2\ MR^2$, $M = 2$ kg, $R = 0.5$ m) spins at $\omega_0 = 2$ rad/s. A piece of clay ($m = 0.1$ kg) travels at $v = 6.5$ m/s and sticks to the edge of the disk. Find the angular velocity after the collision.

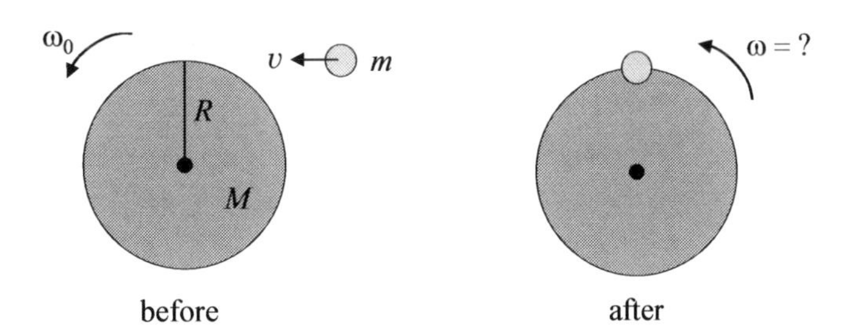

5. A ballerina (I = 20 kg·m^2) is on a raft (I = 100 kg·m^2) on a lake, initially at rest. Suddenly, she spins at 20 rad/s clockwise. Find the angular velocity of the raft. Ignore friction.

Challenging problems

1. When Halley's Comet is furthest from the sun, a distance of 35 A.U., it travels 0.88 km/s. (One A.U. is the distance from Earth to the sun). Its closest distance to the sun is 0.6 A.U. Find its speed at this distance.

2. A 2 kg disk spins at 40 rpm, counterclockwise. A second spinning disk has the same radius but a mass of 4 kg. The second disk lands on top of the first disk, and they spin 40 rpm *clockwise*. What was the rotational speed (rpm) and direction (clockwise or counterclockwise) of the second disk?

3. Lester shoots a wooden wheel ($I = MR^2$, $M = 5$ kg, $R = 0.7$ m) that is free to rotate about its central axis. The 0.1 kg bullet travels horizontally at a speed of 100.0 m/s, 0.3 m above the axis. It passes through the wood and emerges with a horizontal velocity of 95.1 m/s. Find the angular velocity of the wheel.

4. A 4 kg mass is on a frictionless table. It is attached to a 0.05 m diameter pole by a 2 m long rope. The mass travels in a circular orbit at 5 m/s. The rope winds around the pole. After some time, the mass has completed 3 orbits.

   a. Find the speed of the mass at that time.
   b. Find the tension of the string.

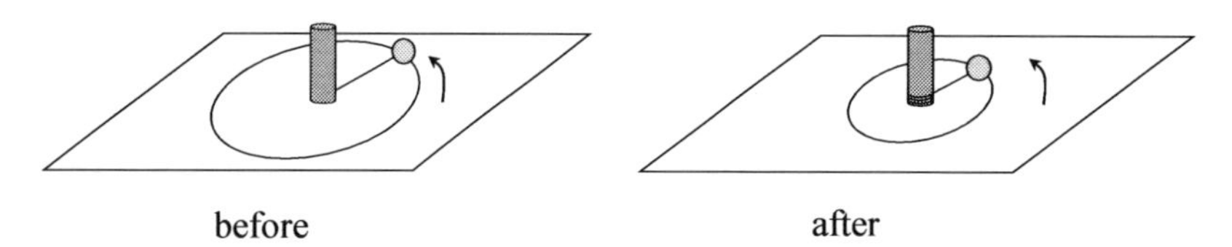

11 Static equilibrium

Center of gravity

Objects have various shapes and sizes. It is useful to define a single point to specify the exact location of an object or collection of objects. This point is called the *center of gravity*, also called the "center of mass." Suppose we have several masses (m_1, m_2, ...). The masses are located at (x_1, y_1), (x_2, y_2), etc. The center of gravity (x_{cg}, y_{cg}) is a weighted average, defined as

$$x_{cg} = \frac{m_1 x_1 + m_2 x_2 + \ldots}{m_1 + m_2 + \ldots}$$

$$y_{cg} = \frac{m_1 y_1 + m_2 y_2 + \ldots}{m_1 + m_2 + \ldots}$$

A heavy mass will tend to bring the center of gravity toward it.

1. A 4 kg mass is connected to a 1 kg mass by a 0.5 m long massless rod. Find the center of gravity (◕).

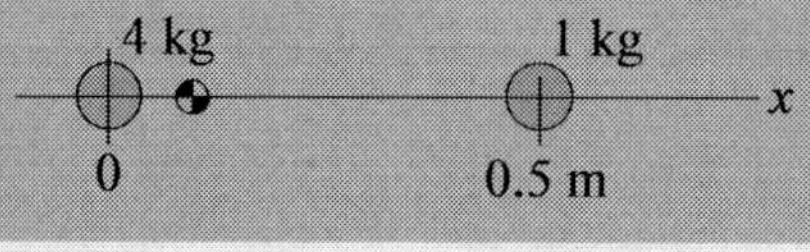

4 kg
1 kg
x
0
0.5 m

$$x_{cg} = \frac{(4)(0) + (1)(0.5)}{4 + 1} = \mathbf{0.1\ m}$$

2. Calculate the center of gravity for these three objects.

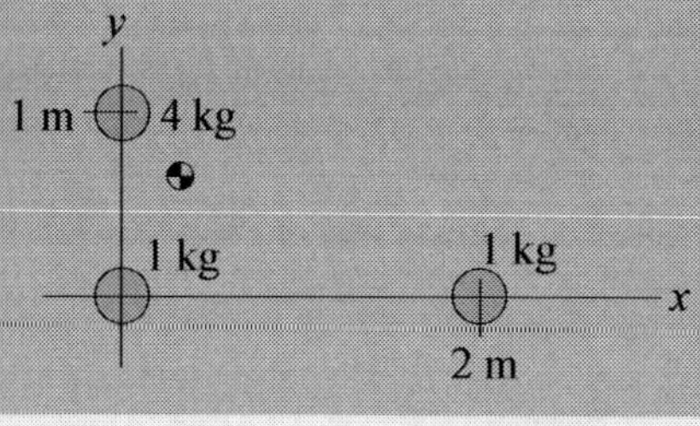

y
1 m
4 kg
1 kg
1 kg
x
2 m

$$x_{cg} = \frac{(4)(0) + (1)(0) + (1)(2)}{4 + 1 + 1} = \mathbf{0.33\ m} \qquad y_{cg} = \frac{(4)(1) + (1)(0) + (1)(0)}{4 + 1 + 1} = \mathbf{0.67\ m}$$

Symmetrical objects

For a symmetrical object like a uniform sphere, cube, rectangle, or cylinder, the center of gravity is simply the center of the object. To find the center of gravity for a collection of objects, first find the center of gravity for each individual object. Then, find the center of gravity for the collection.

1. A mace consists of a 1.5 m long rod (m = 5 kg) and a 0.05 m radius ball (M = 10 kg). Find the center of gravity.

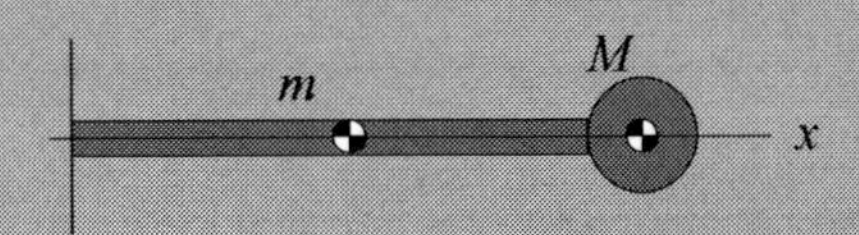

Rod: $m = 5$ kg, $x_{cg} = 0.75$ m (
x_{cg} is half the length of the rod)

Ball: $M = 10$ kg, $x_{cg} = 1.5 + 0.05 = 1.55$ m

$$x_{cg} = \frac{(5)(0.75) + (10)(1.55)}{5 + 10} = \underline{\mathbf{1.28\ m}}$$

2. A uniform cylinder is on top of a uniform cube. How far above the surface is the center of gravity?

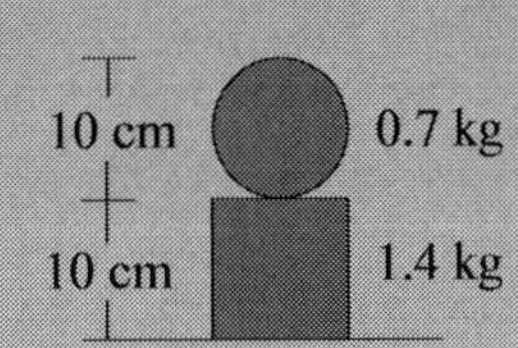

Cube: $m = 1.4$ kg, $y_{cg} = 5$ cm

Cylinder: $m = 0.7$ kg, $y_{cg} = 10$ cm + 5 cm = 15 cm

$$y_{cg} = \frac{(1.4)(5) + (0.7)(15)}{1.4 + 0.7} = \underline{\mathbf{8.3\ cm}}$$

Quiz 11.1

1. A skinny person (50 kg) stands 10 m away from a "big-boned" person (150 kg). How far from the skinny person is the center of gravity?

2. Calculate the center of gravity for these four objects.

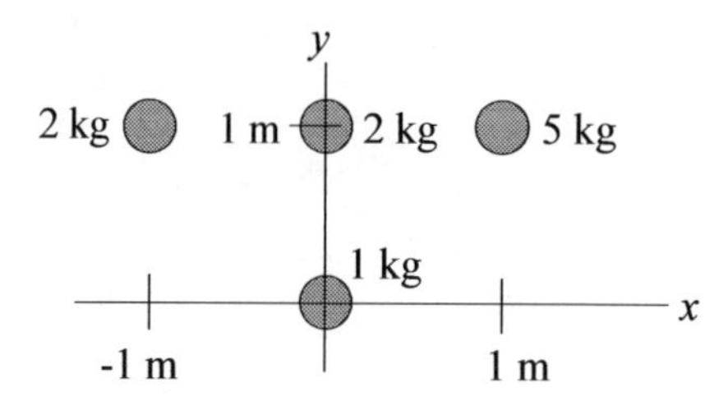

3. The distance from the center of the earth (6×10^{24} kg) to the center of the moon (7×10^{22} kg) is 384,400 km. How far is the center of gravity from Earth's center?

4. A 2 m × 2 m cardboard square has a 1 m × 1 m square section cut out. Where is the center of gravity?

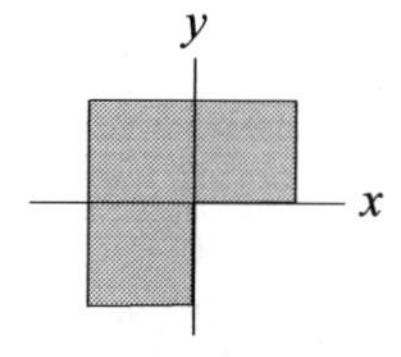

Conditions for static equilibrium

If something is not moving, it is *static*. The sum of the forces is zero. Also, the sum of the torques is zero. We can choose any pivot for calculating torque. From Chapter 9,

$$\tau = rF \sin\theta \quad (\text{if } \theta = 90^\circ, \tau = rF)$$

The gravity force acts on an object's center of mass.

1. A weightlifter pushes on a 1.8 m long massless bar and two 30 kg masses. The left hand is 0.3 m from the left mass. The right hand is 0.2 m from the right mass. Find the force exerted by each hand.

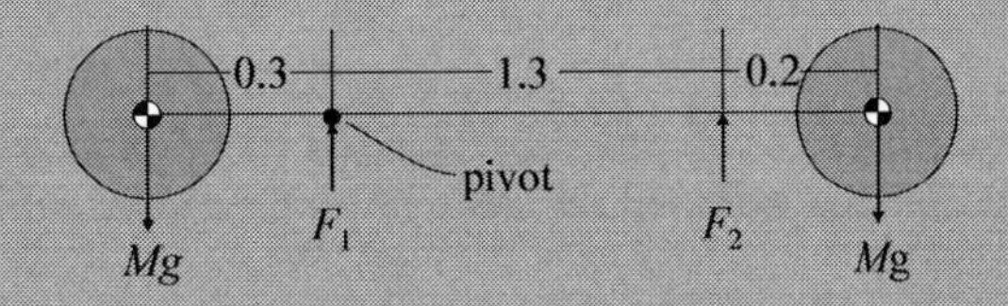

Sum of forces: $-Mg + F_1 + F_2 - Mg = 0$

$F_1 + F_2 = 2Mg = 2(30)(10)$

$\underline{F_1 + F_2 = 600 \text{ N}}$

Sum of torques: $(0.3)Mg + (0)F_1 + (1.3)F_2 - (1.5)Mg = 0$

$(1.3)F_2 = (1.2)Mg$

$F_2 = (1.2)(30)(10)/1.3 =$ **<u>277 N (right hand)</u>**

Plug this into underlined equation: $F_1 + 277 = 600$

$F_1 =$ **<u>323 N (left hand)</u>**

2. A 7 kg horizontal board is 0.5 m long. It is supported by a hinge and rope. Find the tension of the rope.

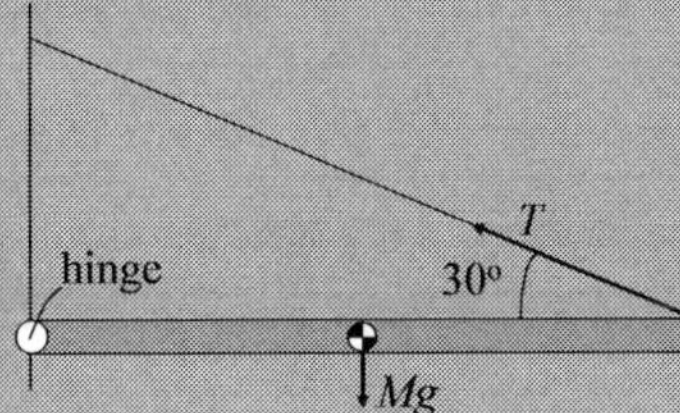

Choose the hinge as the pivot. Then we won't have to worry about the force exerted by the hinge.

Sum of torques: $(0)(F_{\text{hinge}})\sin\theta - (0.25)Mg + (0.5)T\sin(30^\circ) = 0$

$-(0.25)(7)(10) + (0.5)T(0.5) = 0$

$(0.25)T = (0.25)(70)$

$T =$ **<u>70 N</u>**

Moment arm

When calculating torque, **r** points from the pivot to the point where the force is applied. The *moment arm*, or "lever arm," is denoted $r_\perp$. It is the perpendicular distance from the pivot to the line of the force vector.

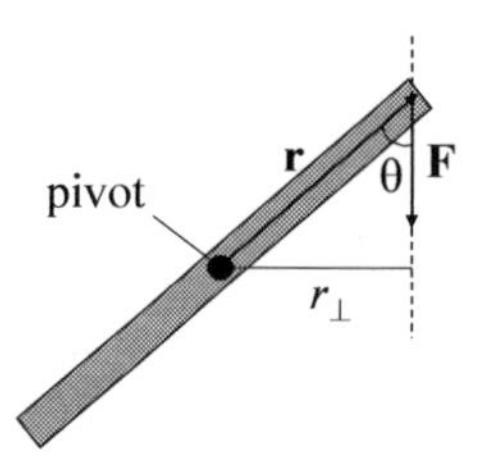

From the definition of torque, $\tau = rF \sin \theta$. From trigonometry, $r \sin \theta = r_\perp$. Hence, an equivalent expression for torque is:

$$\tau = r_\perp F$$

1. A 3 m long beam (40 kg) is supported at each end. A 60 kg object is 1 m from the left support. How much force is exerted by each support?

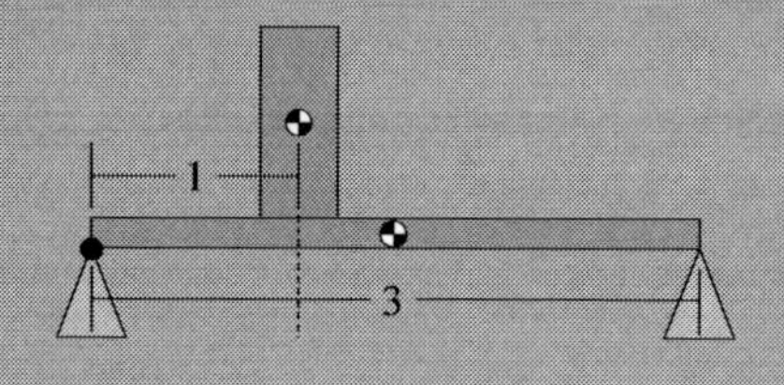

Sum of forces: $F_{left} + F_{right} = (40)(10) + (60)(10)$
$\underline{F_{left} + F_{right} = 1000 \text{ N}}$

Sum of torques: $(0)(F_{left}) - (1)(60)(10) - (1.5)(40)(10) + (3)(F_{right}) = 0$
$-1200 + 3\,F_{right} = 0$
$\underline{\mathbf{F_{right} = 400 \text{ N}}}$

Plug this into underlined equation: $\underline{\mathbf{F_{left} = 600 \text{ N}}}$

2. A woman who weighs 600 N does a "plank pose." Find the normal force exerted by the floor on her hands.

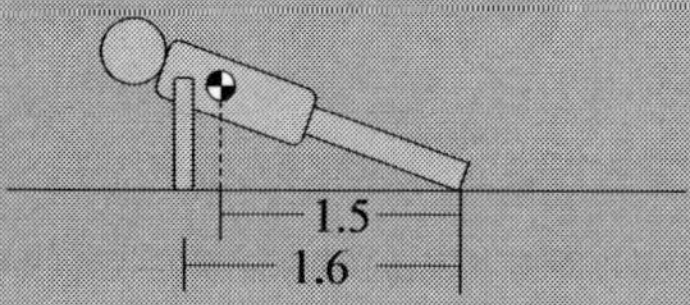

Sum of torques (pivot = feet): $(1.5)(600) - (1.6)n = 0$
$1.6n = 900$, or $n = \underline{\mathbf{562.5 \text{ N}}}$

(Or, 562.5/2 = 281 N per hand)

Quiz 11.2

1. A person pushes on a 1.8 m long 45 kg bar. Find the force exerted by each hand.

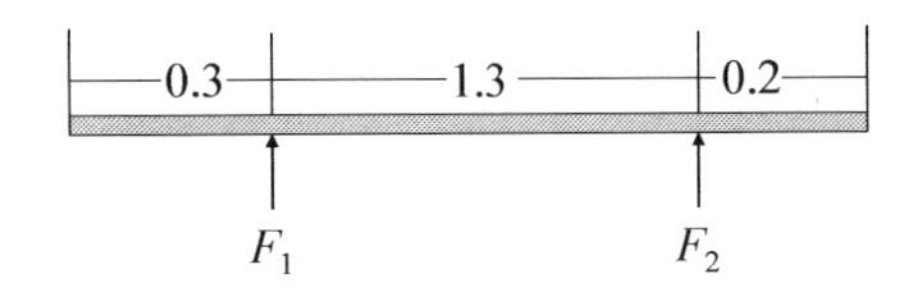

2. A 2 kg shelf is supported by a hinge and rope. A 1 kg object is on the edge. Find the tension of the rope.

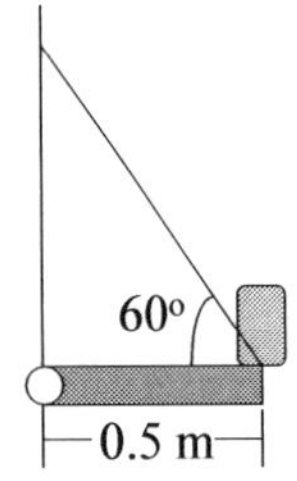

3. A 90 kg person stands at the end of a 3 m long 40 kg diving board. Find the vertical component of the force exerted by the hinge.

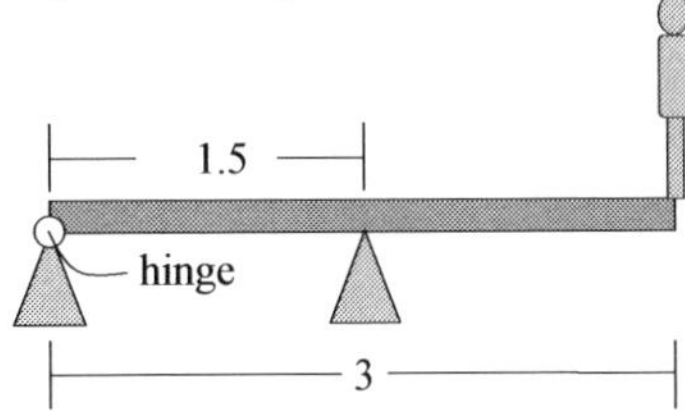

4. A person weighing 500 N does a pushup from the knees. Find the normal force on the hands.

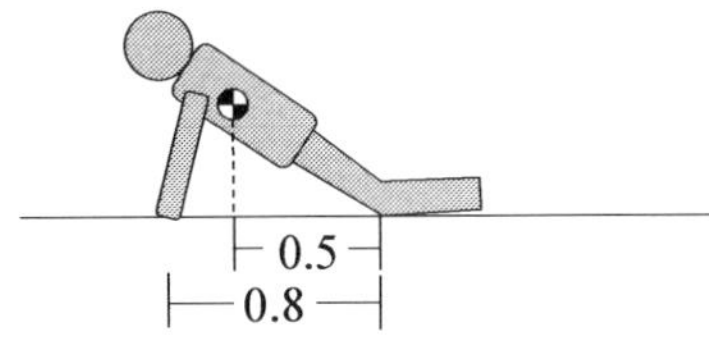

Balance

Objects that are balanced are in static equilibrium. The sum of torques about the pivot equals zero.

1. A dad (M = 100 kg) sits 0.7 m from the pivot of a see-saw. The child sits 2.0 m from the pivot. The see-saw is balanced. What is the child's mass?

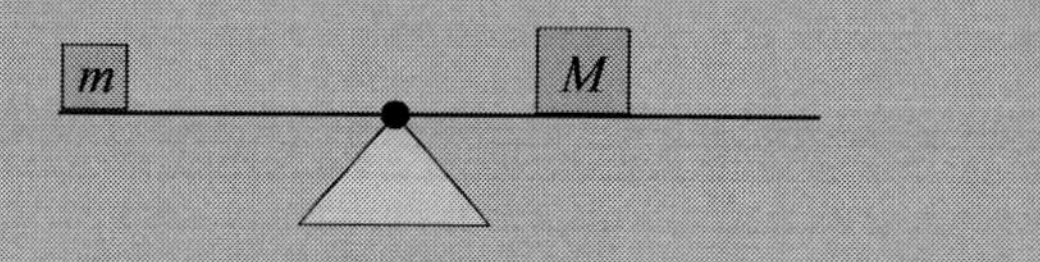

$$\tau = (2.0)mg - (0.7)Mg = 0$$
$$(2)m(10) - (0.7)(100)(10) = 0$$
$$20m = 700$$
$$m = \underline{\mathbf{35\ kg}}$$

2. A mass (m = 2 kg) is placed on a 1.0 m long board (M = 1.5 kg), which sits on a table. How far can the mass be from the edge of the table before the board begins to tilt?

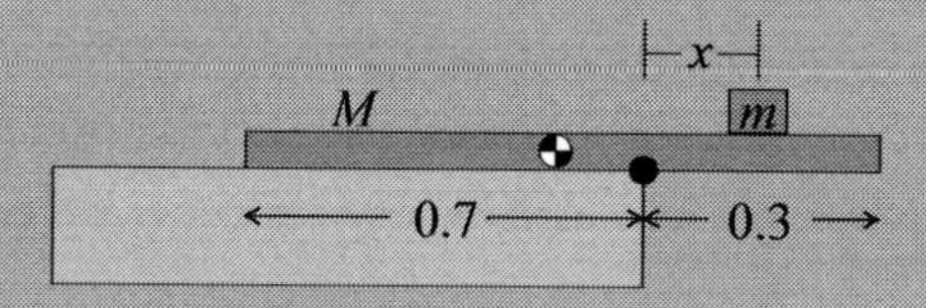

Find the distance x such that the mass and the board's center of mass are perfectly balanced. The pivot is the edge of the table.

$$\tau = (0.2)Mg - xmg = 0$$
$$(0.2)(1.5)(10) - x(2)(10) = 0$$
$$3 = x(20)$$
$$\underline{\mathbf{0.15\ m}} = x$$

Stability

Tall objects tend to tip over more easily than short ones. An object has a *base of support* that is in contact with the ground. When the object rotates, the center of gravity moves. If it doesn't move too far, then gravitational torque will bring it back upright. The *critical angle* is where the center of gravity is directly over the pivot. If the angle exceeds the critical angle, then gravitational torque will cause the object to fall over.

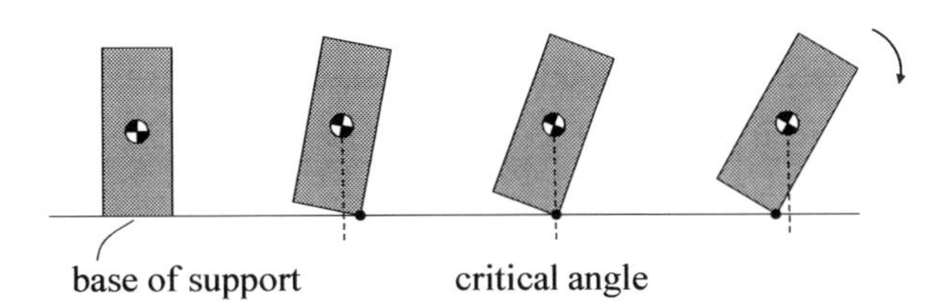

1. Suppose an SUV has a center of gravity 1.4 m above the ground. Its tires are 1.6 m apart. Find the critical angle for tipping.

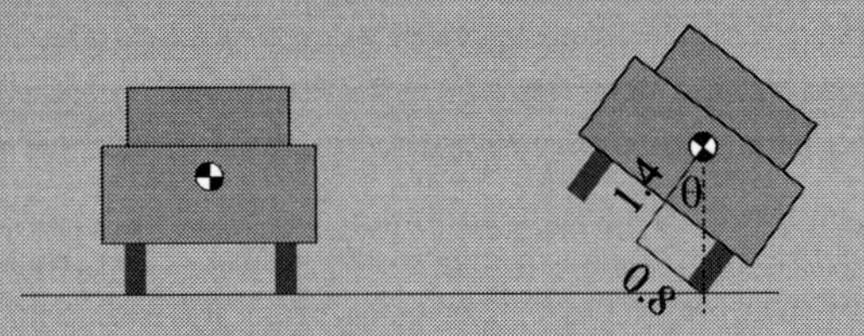

From trigonometry: $\tan\theta = 0.8/1.4 = 0.57$

$\theta = \tan^{-1}(0.57)$

$= \mathbf{\underline{30^\circ}}$

2. A toy model of a leaning tower has a base diameter of 0.1 m and a vertical angle of 22°. How tall can the tower be before it tips over?

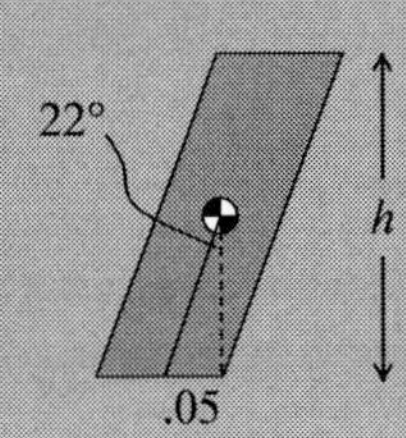

The picture shows the maximum height for the tower. If h is greater than this, then the tower will tip over.

From trigonometry: $\tan(22^\circ) = 0.05/(h/2)$

$0.4 = 0.1/h$

$h = 0.1/0.4 = \mathbf{\underline{0.25\ m}}$

Quiz 11.3

1. A 3 kg mass is on a balance scale, 0.4 m from the pivot. It is balanced by a 1 kg mass. How far is the 1 kg mass from the pivot?

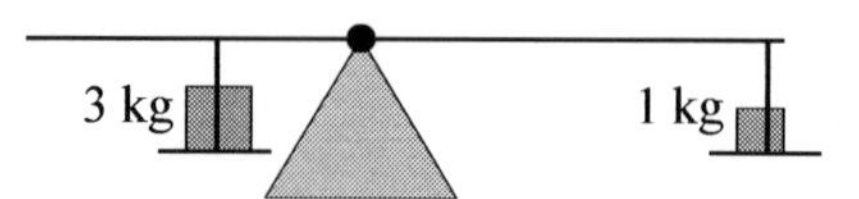

2. A 10 kg board hangs over the roof of a building. A 5 kg cat walks on the board. How far from the right edge of the board is the cat when the board begins to tip?

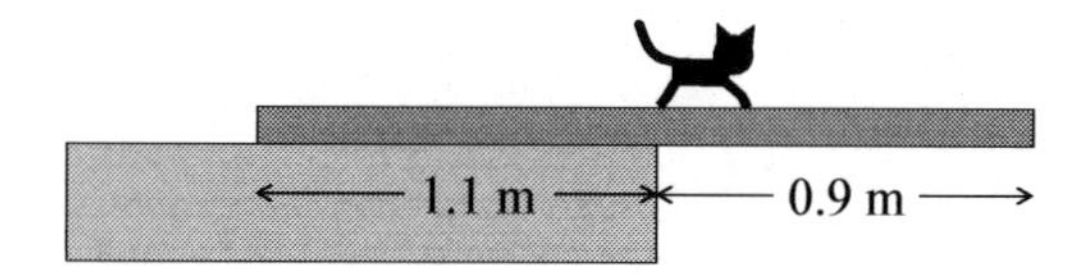

3. Workers want to tip over a crate that has a 2 m × 2 m base and 1.5 m height. To what angle must the crate be rotated?

4. A toy model of a leaning tower has a vertical angle of 12° and a height of 4 m. What is the minimum base diameter for the tower to be stable?

Chapter summary

Center of gravity

$$x_{cg} = \frac{m_1 x_1 + m_2 x_2 + \ldots}{m_1 + m_2 + \ldots}$$

$$y_{cg} = \frac{m_1 y_1 + m_2 y_2 + \ldots}{m_1 + m_2 + \ldots}$$

Static equilibrium:

- Sum of forces equals zero.
- Sum of torques, about any pivot, equals zero.

Torque $\tau = rF \sin\theta = r_{\perp}F$

End-of-chapter questions

1. A 6 kg wood cylinder has a base diameter of 0.1 m and height 1.5 m. A 4 kg metal cube (0.1 m × 0.1 m × 0.1 m) is glued to the top.
   a. How high is the center of gravity?
   b. How far can the object rotate (degrees) before it tips over?

2. Kevin and Rodney hold the ends of a 10 kg, 4 m long board. A 6 kg object sits on the board, 1 m from Kevin. Find the vertical force exerted by Kevin and by Rodney.

3. A 5.0 m long horizontal board has a mass of 4 kg. It is attached to a cable and hinge. Find the tension of the cable.

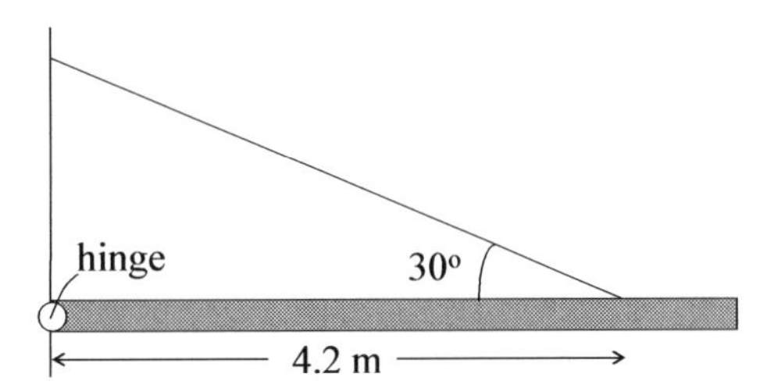

4. A 70 kg person walks on a 35 kg beam that rests on two supports. How close to the right end of the beam can the person get, before it begins to tip?

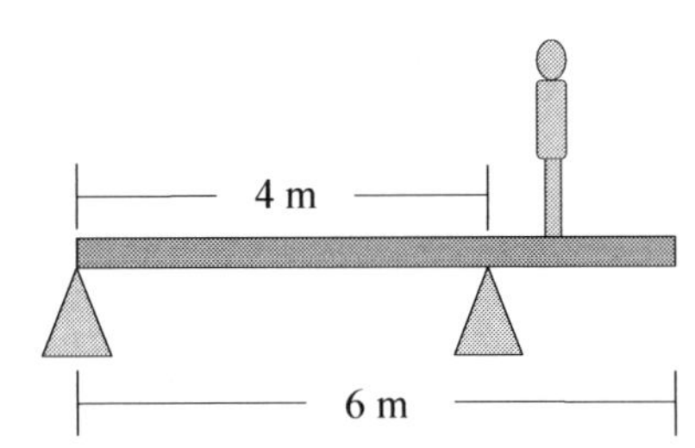

Challenging problems

1. A horizontal board (4 kg) is supported by a rope and hinge.

   a. Find the tension of the rope.
   b. Find the force (F_x and F_y) exerted by the hinge.

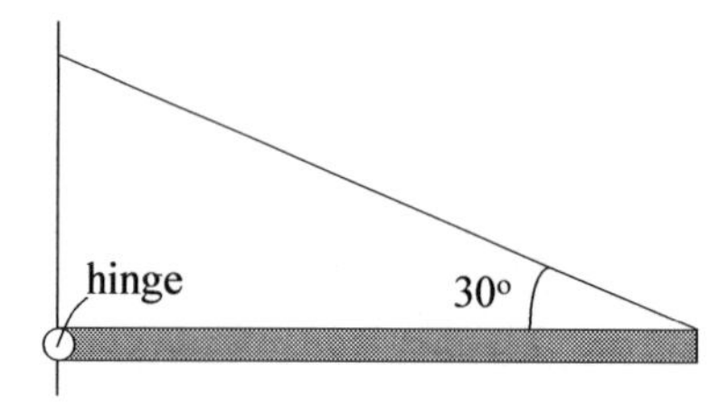

2. A 2 m long, 1.5 kg board is at horizontal angle of 40°. It is supported by a rope and hinge. Find the tension of the rope.

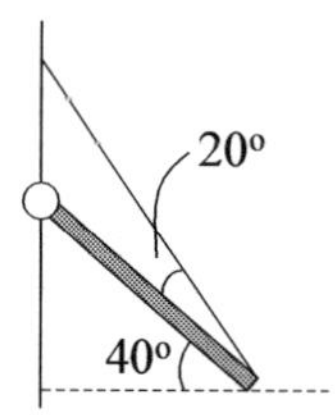

3. A 20 kg ladder rests against a frictionless wall. (The floor does have friction). Find the normal force exerted by the wall.

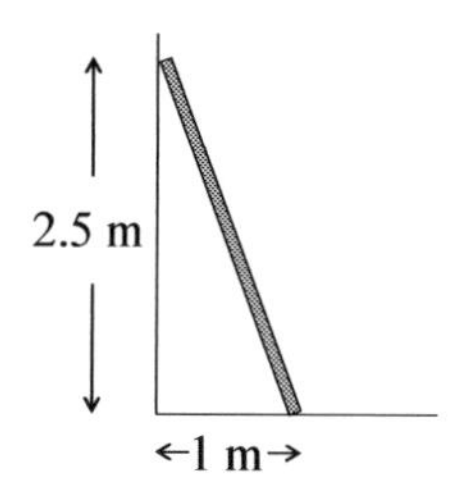

4. Find the maximum value of x such that the object will not tip over.

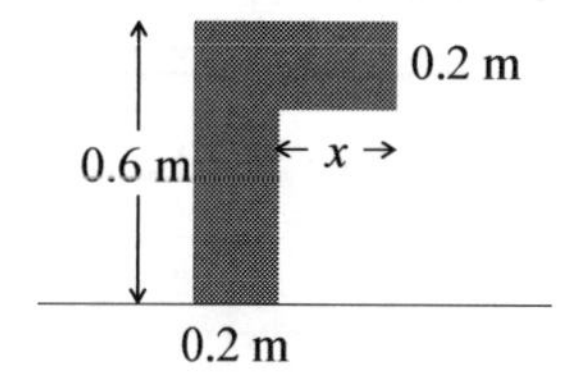

12 Oscillations

Hooke's law

Consider a mass attached to a spring. The spring is relaxed, not stretched or compressed. The mass is at its *equilibrium* position, denoted $x = 0$. A person pulls on the mass so that its position is $x > 0$. The spring is now stretched.

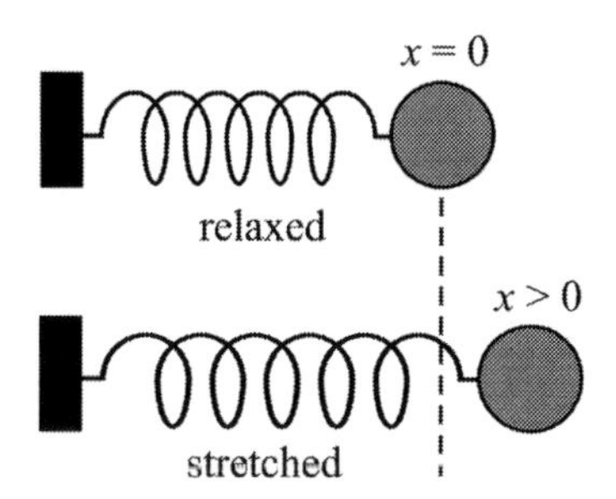

According to Hooke's law, the spring exerts a force on the mass,

$$F = -kx$$

where k is the *spring constant*. The minus sign means if the mass has a positive x value, the spring will pull in the $-x$ direction. If the mass has a negative x value, the spring will push in the $+x$ direction.

1. A person pulls a spring with a 10 N force, causing it to stretch 0.1 m in the x direction. What is the spring constant?

The spring exerts a force of 10 N in the $-x$ direction: $F = -kx$

$$-10 = -k(0.1)$$

$$\underline{\textbf{100 N/m}} = k$$

2. A spring ($k = 50$ N/m) hangs vertically. A 2 kg mass is attached. How much does the spring stretch?

Gravity pulls down with a force: $mg = (2)(10) = 20$ N
The spring pulls up with a force: $kx = (50)x$

These forces must balance: $20 = 50x$

$$\underline{\textbf{0.4 m}} = x$$

(Notice how we neglected the minus signs. This is OK so long as you think about what direction the forces are acting.)

Oscillating spring and mass

Suppose a person pulls the mass to $x = A$. Then, at $t = 0$, the person lets go. It will *oscillate* back and forth. In one cycle, the mass goes from A to $-A$ and back to A. (The dashed line is $x = 0$).

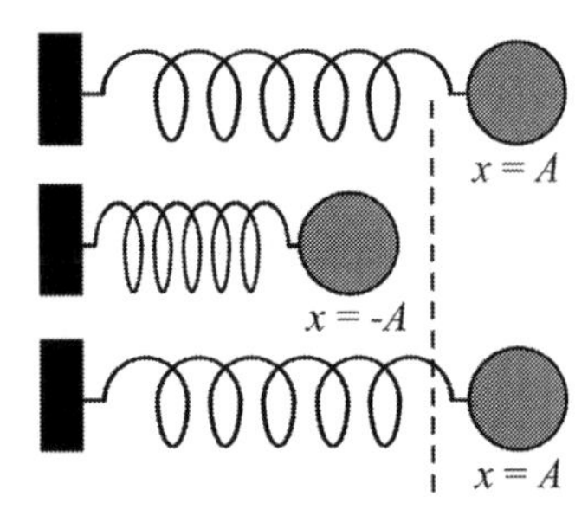

A is the *amplitude* in meters (m). ω is the *angular frequency* (rad/s) and tells how rapidly the mass oscillates. For a spring and mass,

$$\omega = \sqrt{k/m}$$

The *frequency* (f) is the number of times the mass goes back and forth each second:

$$f = \omega/(2\pi)$$

Frequency has units of s^{-1}, also called Hertz (Hz) or cycles per second. The *period* (T) is the time it takes (s) to complete one cycle:

$$T = 1/f$$

1. A 4 kg mass is attached to a spring, $k = 16$ N/m. It is displaced 5 cm from equilibrium and released. Find the amplitude, frequency, and period of its oscillation.

$A = 5$ cm = **0.05 m**

$\omega = \sqrt{16/4} = 2$ rad/s

$f = 2/(2\pi) =$ **0.32 s^{-1}**

$T = 1/f =$ **3.1 s**

2. A 0.1 kg mass hangs vertically from a spring with a spring constant 10 N/m. A person pulls it down 2 cm and lets go. How long does it take to reach the maximum height?

$\omega = \sqrt{10/0.1} = 10$ rad/s

$f = 10/(2\pi) = 1.6$ s^{-1}

$T = 1/f = 0.6$ s

The mass goes from $y = -2$ cm to $y = 2$ cm. The time it takes is half the period.

$T/2 =$ **0.3 s**

Quiz 12.1

1. A mass is attached to a spring, $k = 17$ N/m. The mass is pushed 0.3 m in the $-x$ direction, compressing the spring. Find the force exerted by the spring on the mass.

2. A 1 kg mass hangs vertically from a spring. A 2 kg mass is attached to the 1 kg mass. This causes the spring to stretch by an additional 0.1 m. Find the spring constant.

3. A 10 g mass is attached to a spring with a spring constant 10 N/m. It is displaced 7 cm from equilibrium and released. Find the amplitude, frequency, and period of the oscillation.

4. A 2 kg mass hangs vertically from a spring, $k = 20$ N/m. The professor raises the mass 5 cm and releases it. How long does it take to reach the equilibrium position ($y = 0$)?

Velocity and acceleration

Suppose the mass is stretched to $x = A$ and released at $t = 0$. Its position (x) for $t > 0$ is given by a cosine function. Its velocity (v) and acceleration (a) are described similarly.

$x = A\cos(\omega t)$

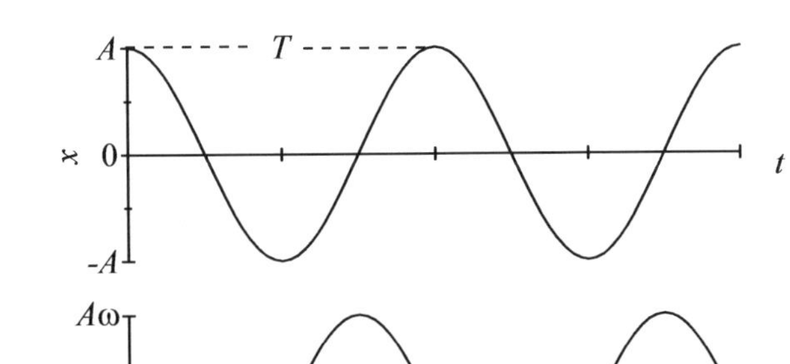

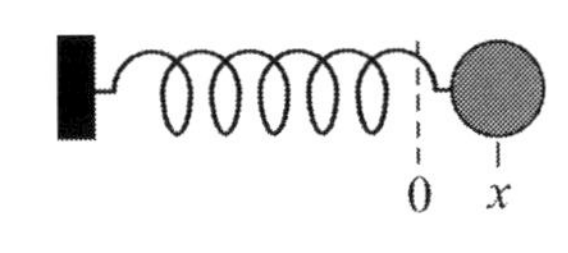

$v = -A\omega\sin(\omega t)$

$a = -A\omega^2\cos(\omega t)$

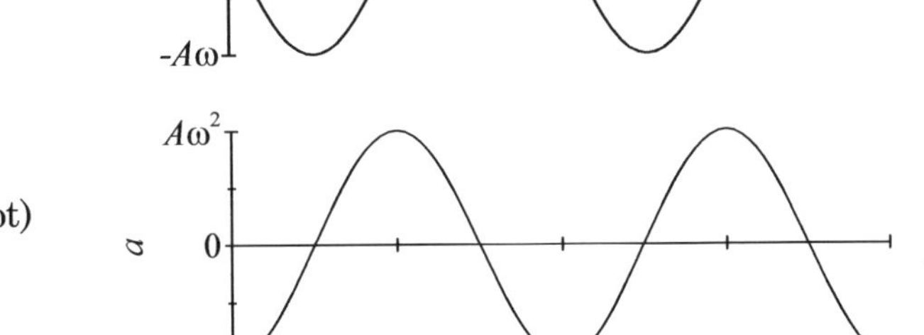

1. A 0.1 kg mass is attached to a spring, k = 10 N/m. It is displaced 0.2 m from equilibrium and released. What is its maximum speed and where does it occur?

$v = -A\omega\sin(\omega t)$
Since sin is between −1 and 1, the maximum speed is $\underline{v_{max} = A\omega}$

$\omega = \sqrt{10/0.1} = 10$ rad/s
$v_{max} = (0.2)(10) = \underline{\textbf{2 m/s}}$

From the graphs, the maximum speed occurs when $\underline{\boldsymbol{x = 0}}$ (equilibrium).

2. In the previous problem, what is the maximum acceleration magnitude, and where does it occur?

$a = -A\omega^2\cos(\omega t)$
Since cos is between −1 and 1, the maximum acceleration is $A\omega^2 = (0.2)(100) = \underline{\textbf{20 m/s}^2}$

The maximum acceleration magnitude occurs when $\underline{\boldsymbol{x = -A \text{ or } A}}$ (when the spring is maximally compressed or stretched).

$\underline{\boldsymbol{x = -0.2 \text{ m or } 0.2 \text{ m}}}$

Energy

As we saw in Chapter 7, kinetic energy is $K = \frac{1}{2}mv^2$. For a spring, the potential energy is

$$U = \tfrac{1}{2}kx^2$$

When $x = 0$, the spring is relaxed and $U = 0$. If the mass is at $x = A$, then $U = \frac{1}{2}kA^2$. If $x = -A$, U is also $\frac{1}{2}kA^2$. The minus sign disappears because x is squared.

1. A 0.05 kg mass is attached to a spring, $k = 40$ N/m. The spring is compressed by 10 cm. Find the potential energy.

$U = \frac{1}{2}(40)(.10^2) = \underline{\mathbf{0.2\ J}}$

2. In the previous problem, the mass is released. Use energy conservation to find its kinetic energy at equilibrium ($x = 0$).

$E = K + U$
$E(\text{before}) = U = 0.2$ J (All potential)
$E(\text{after}) = K$ (All kinetic)

$E(\text{before}) = E(\text{after})$
$\underline{\mathbf{0.2\ J}} = K$

3. In the previous problem, find the speed at equilibrium.

$K = \frac{1}{2}mv^2$
$0.2 = \frac{1}{2}(0.05)v^2$
$8 = v^2$
$\underline{\mathbf{2.8\ m/s}} = v$

A second way to solve this problem: $\omega = \sqrt{40/0.05} = 28$ rad/s

$v_{max} = A\omega$
$= (0.1)(28) = \underline{\mathbf{2.8\ m/s}}$

Quiz 12.2

1. A 0.4 kg mass is attached to a spring, $k = 10$ N/m. It is displaced 5 cm from equilibrium and released. What is its speed at equilibrium ($x = 0$)?

2. In the previous problem, find the acceleration of the mass when the spring is maximally compressed.

3. A 5 kg mass is attached to spring, $k = 0.2$ N/m. It oscillates with amplitude 0.1 m. Find the potential energy when the spring is maximally stretched.

4. In the previous problem, what is the kinetic energy at equilibrium?

Swinging pendulum

A simple pendulum is a small mass attached to a string. The mass is free to swing back and forth.

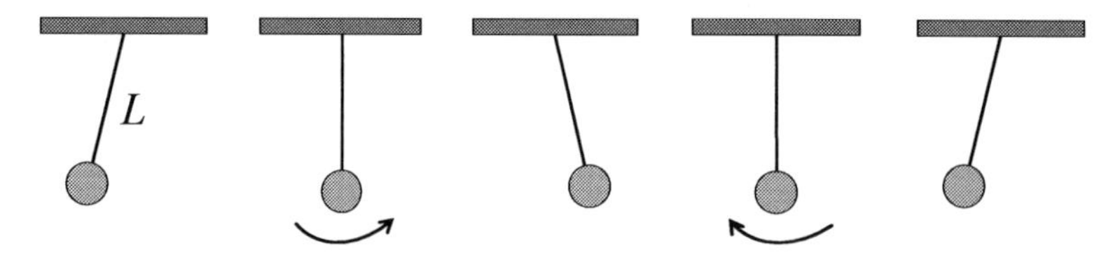

The angular frequency of the oscillation is

$$\omega = \sqrt{g/L}$$

where L is the length of the string. A longer string means a lower frequency. Note that the frequency does not depend on mass.

1. A mass hangs from a 10 cm string and swings back and forth. Find the frequency and period of the oscillation.

$\omega = \sqrt{10/0.1} = 10$ rad/s

$f = 10/(2\pi) =$ **1.6 s^{-1}**

$T = 1/f =$ **0.6 s**

2. A bowling ball hangs from a 10 m rope. A professor displaces the ball and then releases it. How long does it take for the ball to reach the bottom of the swing?

$\omega = \sqrt{10/10} = 1.0$ rad/s

$f = 1.0/(2\pi) = 0.16$ s^{-1}
$T = 1/f = 6$ s

Going from the top to the bottom of the swing is ¼ of a cycle.

$T/4 = 6/4 =$ **1.5 s**

Damped and driven oscillations

As we saw previously, the position of an oscillating mass (ignoring friction) is given by

$$x = A\cos(\omega t)$$

If there is friction, the oscillation amplitude will decrease steadily over time. This is called a *damped* oscillation. The amplitude decays exponentially:

$$x = (Ae^{-t/\tau})\cos(\omega t)$$

where τ is the *time constant.*

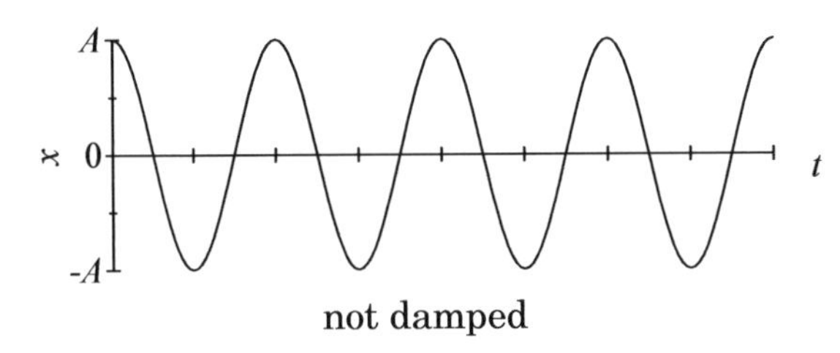

not damped

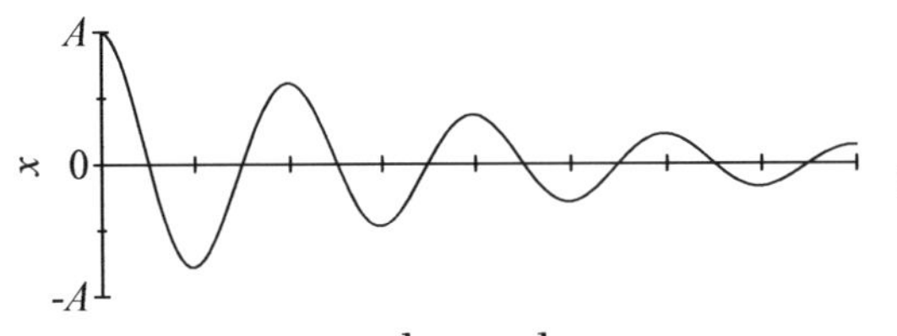

damped

A *driven* oscillation occurs when a force pushes or pulls on the mass with a regular frequency. The amplitude is maximized when the driving frequency equals the *natural frequency* of the oscillator. This is just the frequency when there is no driving force.

For a spring and mass, the natural frequency is $\omega = \sqrt{k/m}$. For a pendulum, it is $\omega = \sqrt{g/L}$.

1. After 15 s, the amplitude of an oscillator is 40% its original value. What is the time constant?

The amplitude is $Ae^{-t/\tau}$. At $t = 15$ s, it equals $0.40A$.

$Ae^{-15/\tau} = 0.40A$
$e^{-15/\tau} = 0.40$
$\ln(e^{-15/\tau}) = \ln(0.40)$
$-15/\tau = -0.916$
$15 = 0.916\,\tau$
16.4 s $= \tau$

2. A girl sits on a swing with a 3 m rope. Her father wants to maximize the amplitude of the swing. With what frequency (s^{-1}) should he push?

$\omega = \sqrt{g/L} = \sqrt{10/3} = 1.8$ rad/s

$f = 1.8/(2\pi) =$ **0.3 s⁻¹**

Quiz 12.3

1. A pendulum consists of a mass connected to a massless rod that is 1 m long. What is the frequency and period of its oscillation?

2. A ball is held by a 40 m long cable. The ball is displaced toward the left and then released. It swings to the right. How long does it take to reach the rightmost point of its swing?

3. A mass oscillates with an amplitude of 5 cm. 20 s later, the amplitude is 1 cm. Find the time constant.

4. A 0.1 kg mass hangs from a spring. The spring constant is 10 N/m. A student taps on the mass with a pencil, causing it to oscillate. If the student wants to maximize the amplitude, how many times should she tap per second?

Chapter summary

Hooke's law	$F = -kx$
Angular frequency (rad/s) for spring and mass	$\omega = \sqrt{k/m}$
Angular frequency (rad/s) for pendulum	$\omega = \sqrt{g/L}$
Frequency (s^{-1})	$f = \omega/(2\pi)$
Period (s)	$T = 1/f$
Position	$x = A\cos(\omega t)$
Velocity	$v = -A\omega\sin(\omega t)$
Acceleration	$a = -A\omega^2\cos(\omega t)$
Maximum speed	$v_{max} = A\omega$
Potential energy for a spring	$U = \frac{1}{2}kx^2$
Damped oscillation	$x = (Ae^{-t/\tau})\cos(\omega t)$

End-of-chapter questions

1. A 0.3 kg mass is attached to a spring, k = 7 N/m. Find the frequency and period of oscillation.

2. A spring is attached to the ceiling and hangs vertically. When a 1 kg mass is attached, the spring stretches by 0.5 m. The mass then undergoes oscillations. Find the frequency.

3. On the moon, $g = 1.6$ m/s^2. An astronaut has a simple pendulum of length 0.5 m. Find the period of oscillation.

4. A 0.1 kg mass is attached to a spring. The spring constant is 10 N/m. The mass oscillates with an amplitude of 7 cm. What is the maximum speed of the mass, and where does this occur?

5. In the previous problem, what is the maximum potential energy, and where does it occur?

6. A damped oscillator's amplitude decreases by 50% in 8 s. What is the time constant?

7. A bowling ball hangs by a 10 m rope. A person pushes the ball at a regular frequency. What frequency will maximize the ball's swing?

Challenging problems

1. A student determines that it takes 5 N of force to stretch a spring by 1 cm. The student then hangs the spring vertically. She attaches a 20 kg mass to the spring.

    a. How much did the spring stretch when the mass was attached?
    b. The student taps the mass at a regular frequency, causing the mass to oscillate up and down. How many taps per second maximize the amplitude?
    c. The mass oscillates with an amplitude of 7 cm. What is its maximum speed, and where does this occur?

2. A 1 kg mass is attached to a relaxed horizontal spring, $k = 4$ N/m. A 3 kg mass collides with the 1 kg mass and sticks to it. How long does it take for the spring to become fully compressed?

3. A toy gun has a spring ($k = 10$ N/m) that is compressed by 10 cm.

    a. What is the potential energy of the spring?
    b. The gun points horizontally and the spring is released, launching a 0.1 kg projectile. Use energy conservation to find the speed of the projectile as it exits the gun.

4. A 30 kg block of wood is suspended by a 2 m long rope. A 0.2 kg projectile travels at 300 m/s and embeds itself in the block. The block swings up to a maximum height of h.

    a. Find h.
    b. How long does it take to reach this height?

13 Waves

Traveling waves

Consider a long string under tension. At the left end, someone wiggles the string up and down. This causes a *transverse wave* to travel toward the right. It is called "transverse" because the atoms of the string oscillate vertically, while the wave travels horizontally. The picture shows a snapshot of the wave. A movie would show the wave pattern moving toward the right.

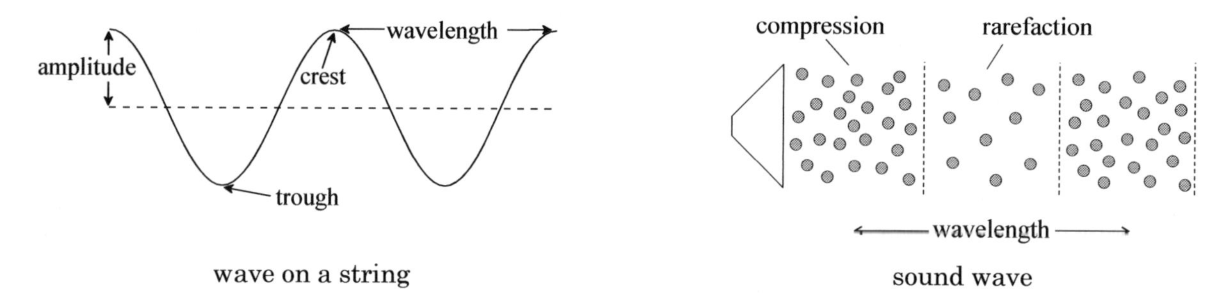

wave on a string

sound wave

Now, consider a speaker membrane that pushes on the air. It causes the air to be compressed to a high pressure. When the speaker membrane moves back, the pressure decreases. A sound wave is created, with regions of high pressure (*compression*) and low pressure (*rarefaction*). This is a *longitudinal wave*. The air molecules move along the same axis as the wave. In the picture, the sound wave travels toward the right.

The maximum of a wave is a *crest* and the minimum is a *trough*. The height of the crest is the *amplitude*. The distance from crest to crest is the *wavelength*. The wavelength is the distance over which the wave pattern repeats.

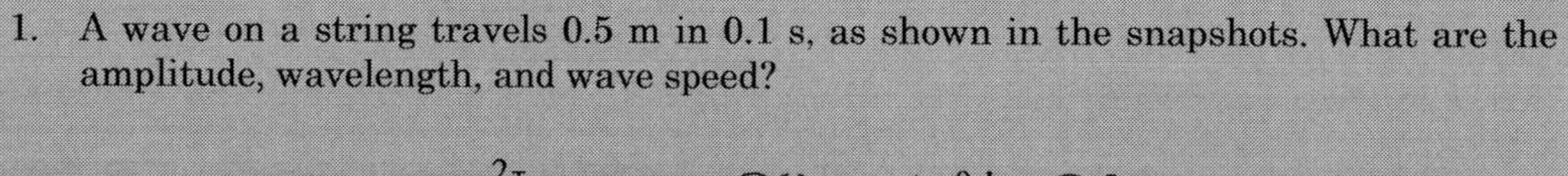

1. A wave on a string travels 0.5 m in 0.1 s, as shown in the snapshots. What are the amplitude, wavelength, and wave speed?

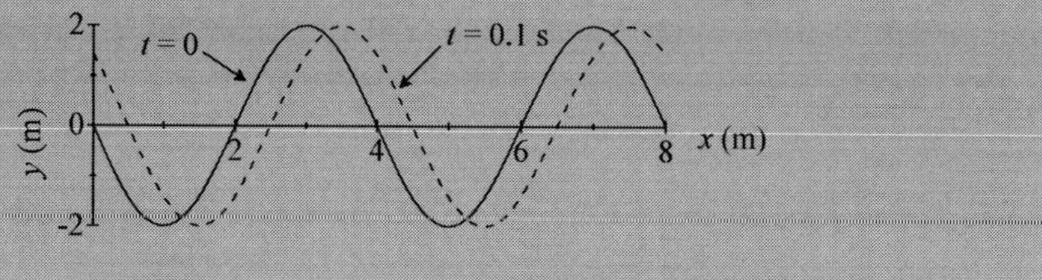

Amplitude = **2 m**

Wavelength = **4 m**

Wave speed = 0.5 m / 0.1 s = **5 m/s**

Sinusoidal waves

A *sinusoidal* wave is given by a sine or cosine function. A cosine wave is

$$y = A\cos(kx \pm \omega t)$$

where A is the amplitude. At $t = 0$, a snapshot of the wave looks like this:

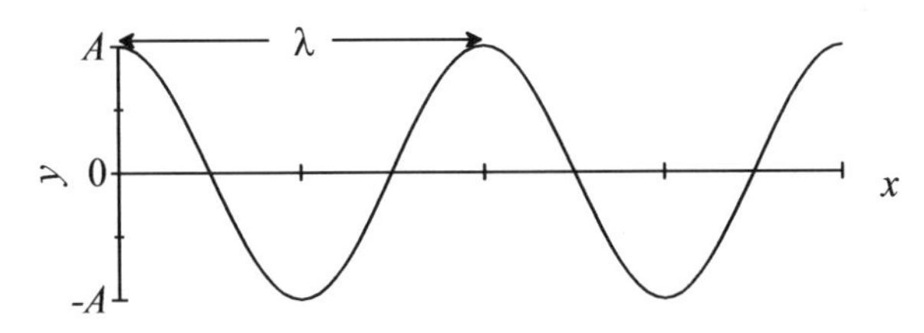

The – is for waves that travel right and + is for waves that travel left. The wavelength (λ) is

$$\lambda = 2\pi/k$$

The *frequency* (f) gives the number of crests that go by each second:

$$f = \omega/(2\pi)$$

The *wave speed* (v) is

$$v = \lambda f$$

1. A sound wave is given by $y = A\cos(6x - 2058t)$, where x is in m and t is in s. Find the wavelength, frequency, wave speed, and direction.

$\lambda = 2\pi/6 =$ **1.05 m**
$f = 2058/2\pi =$ **327.5 s^{-1}**
$v = (1.05)(327.5) =$ **344 m/s**
The – means the wave travels **right (+x direction)**

2. At $t = 0$, a duck floats on the crest of a wave with an amplitude 0.1 m. The wavelength is 2 m and the wave speed is 4 m/s. What is the height (y) of the duck at $t = 1.3$ s?

At $t = 0$, a crest of the wave is at $x = 0$. So, the duck sits at $x = 0$.

For $t > 0$, the duck's height will be $y = (0.1)\cos(0 \pm \omega t) = (0.1)\cos(\omega t)$

$v = \lambda f$
$4 = (2)f$, or $f = 2\ s^{-1}$
$\omega = 2\pi f = 12.6$ rad/s

$y = (0.1)\cos[(12.6)(1.3)] =$ **-0.08 m**

Quiz 13.1

1. A snapshot of a water wave is shown. Find the amplitude and wavelength.

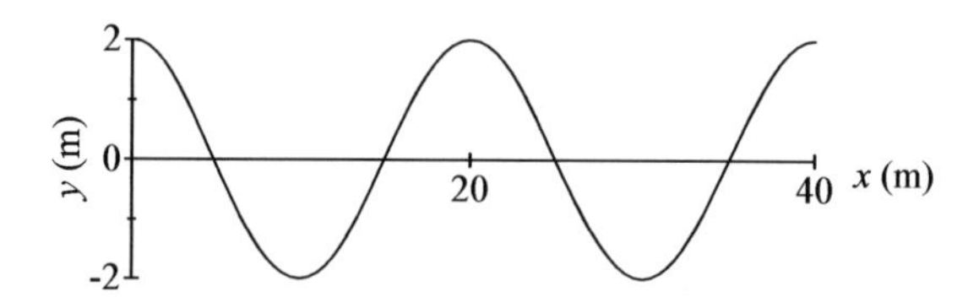

2. In the previous problem, suppose the wave travels toward the right at 4 m/s. Write an equation for the wave.

3. A wave on a string is given by $y = (0.1 \text{ m}) \sin(3x + 6t)$, where x is in m and t is in s. Find the amplitude, wavelength, frequency, wave speed, and direction.

4. A wave on a string has a wavelength 1.5 m, wave speed 10 m/s, and amplitude 0.2 m. A red dot is painted on the string. At $t = 0$, the dot is on the crest of the wave. Find the height of the dot at $t = 0.7$ s.

Standing waves on a string

Consider a stringed instrument like a violin or guitar. The string is clamped, or fixed, at two ends. A vibration of the string is a *standing wave*. The two ends are *nodes*, points where the string does not move. The sine wave equals zero at the nodes. An *antinode* is a point of maximum displacement.

A specific type of standing wave is a *mode*. The longest-wavelength mode is the *fundamental*. For our string, it's half a sine wave. The wavelength is therefore $2L$. The figure shows a snapshot of the string (solid curve). At a later time, it is inverted (dashed curve). The string oscillates between these two profiles.

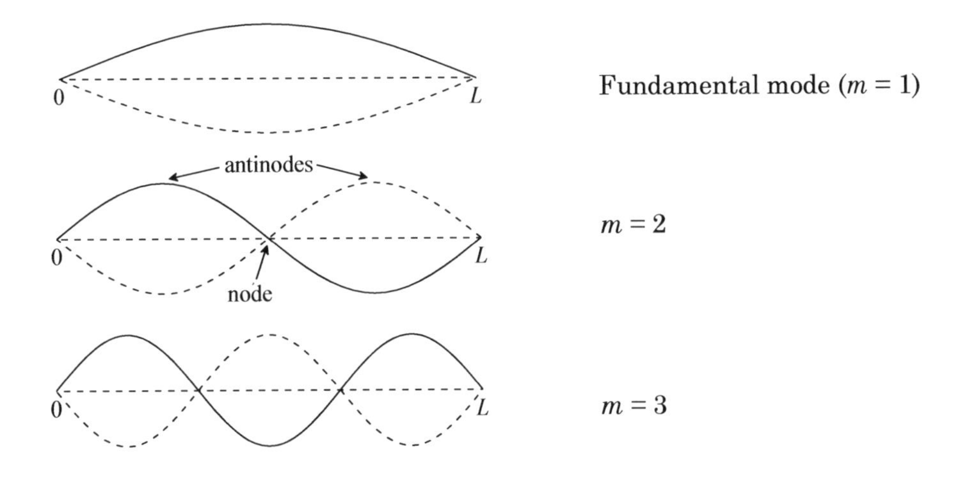

The fundamental is called the $m = 1$ mode. The $m = 2$ mode is a sine wave with a wavelength L. It has a node in the center. The $m = 3$ mode has an additional node and has a wavelength $2L/3$.

1. An instrument has a 1 m long string that has a fundamental mode of 440 Hz. What is the wave speed (the speed of a traveling wave)?

Wavelength of the standing wave: $\lambda = 2L = 2$ m
Frequency of the standing wave: $f = 440$ Hz (or s^{-1})

The speed of a travelling wave on this string would be $v = \lambda f$,
$= (2)(440) =$ **880 m/s**

2. The wave speed on a string is 12 m/s. The string is clamped at two points, 3 m apart. Find the frequencies of the first three modes.

$\lambda = 2L$, L, and $2L/3$
$= 6$, 3, and 2 m

$v = \lambda f$, or $f = v/\lambda$
$= 12/6$, $12/3$, and $12/2$
= **2, 4, and 6 s^{-1} (or Hz)**

Standing sound waves

Air in a tube can produce standing sound waves, where the air molecules move back and forth along the tube axis. Let ΔP denote the difference between the pressure in the tube and that of the air outside the tube. For regions of compression, $\Delta P > 0$. For rarefaction, $\Delta P < 0$.

We plot ΔP superimposed over a picture of the tube. At the ends of the tube, follow these rules:

- If the end is open to the air, then it is a node ($\Delta P = 0$). The outside atmosphere fixes the pressure.
- If the end is closed, then it is an antinode (ΔP is a maximum or minimum).

Here are the fundamental (longest-wavelength) modes for three different tubes:

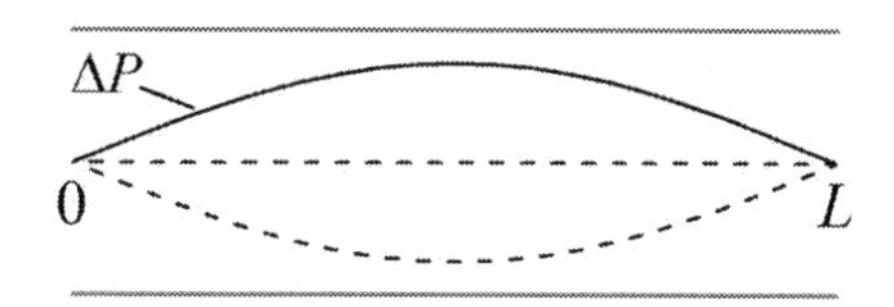

Both ends open

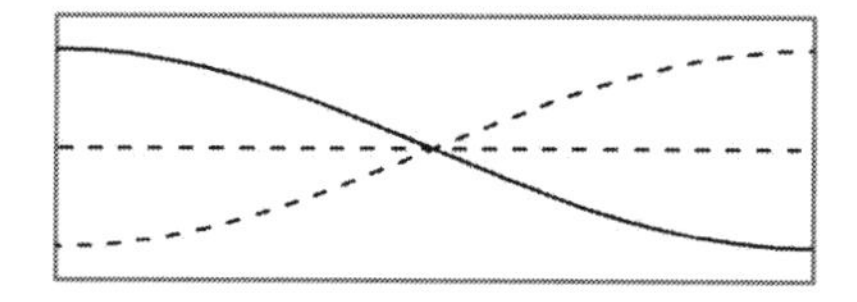

Both ends closed

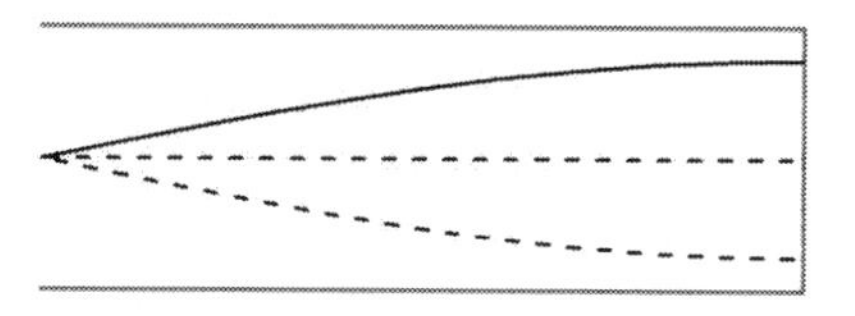

One end open, one end closed

1. Find the wavelengths for the fundamental modes shown in the figures, for a tube length of 0.5 m.

Both ends open: $\lambda = 2L =$ **1 m**

Both ends closed: $\lambda = 2L =$ **1 m**

One end open, one end closed: $\lambda = 4L =$ **2 m**

2. In the previous problem, find the frequencies. The speed of sound in air is 343 m/s.

$v = \lambda f$, or $f = v/\lambda$

Both ends open: $f = 343/1 =$ **343 Hz**

Both ends closed: $f = 343/1 =$ **343 Hz**

One end open, one end closed: $f = 343/2 =$ **172 Hz**

Quiz 13.2

1. A 2 m long string is clamped at both ends. Its fundamental mode has a frequency 300 Hz. What is the wave speed?

2. An instrument has a 0.4 m long string. The wave speed is 200 m/s. Find the frequencies of the first three modes.

3. A 0.3 m long tube is open at both ends. Find the frequency of the fundamental mode.

4. In the previous problem, a cap is placed on one end of the tube. Now what is the frequency of the fundamental mode?

Superposition

When two waves overlap each other, they add together. This is called *superposition*. For example, consider two sound waves of slightly different frequencies (f_1 and f_2).

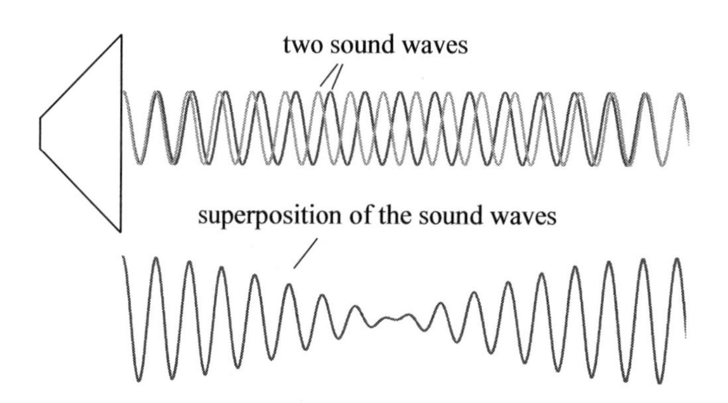

In certain regions, the sine waves add together to make a large-amplitude (loud) wave. In other regions, the sine waves nearly cancel, resulting in a quiet sound. The listener hears this as a loud-quiet-loud-quiet pattern. The *beat frequency* of this pattern is just the difference between the two frequencies,

$$f_{\text{beat}} = |f_1 - f_2|$$

Superposition is important in music. The standing-wave modes (m = 1, 2, 3...) add together to give an instrument its unique sound.

1. A tuning fork emits a 440 Hz sound wave. A second fork has a small weight attached and has a frequency of 437 Hz. What is the beat frequency?

$f_{\text{beat}} = 440 - 437 =$ **<u>3 Hz</u>** (3 beats per second)

2. In the previous problem, find the wavelengths of the sound waves from each tuning fork. The speed of sound in air is 343 m/s.

$v = \lambda f$, or $\lambda = v/f$

$\lambda = 343/440,\ 343/437$

$=$ **<u>0.780 m, 0.785 m</u>**

Interference

Consider two speakers, on the x axis, that emit identical sound waves of the same frequency. A listener is also on the x axis. The wave from each speaker travels a different distance, or path length.

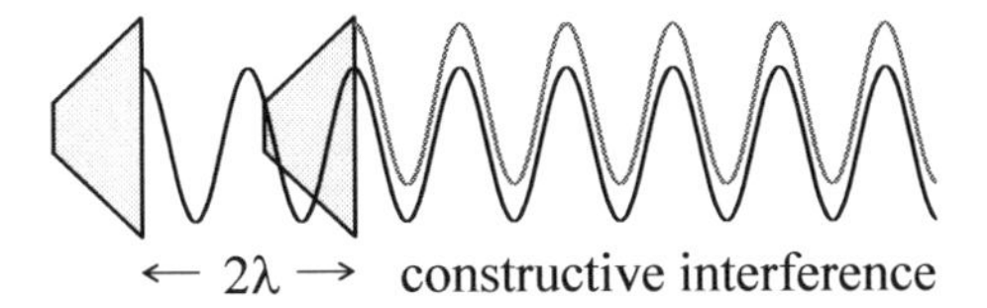

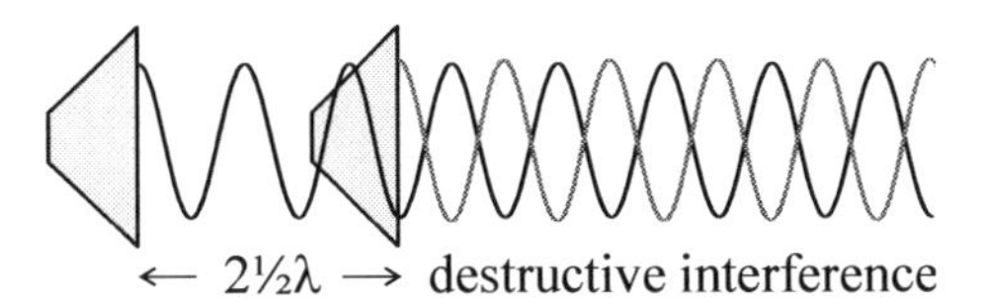

If the path-length difference is a multiple of λ (0, λ, 2λ...), then we have *constructive interference.* The crests of one wave line up with the crests of the other wave. The troughs line up with the troughs. The superposition of the two waves is a big wave. The listener hears a loud sound.

Suppose the path-length difference is a multiple of λ, plus ½λ (½λ, 1½λ, 2½λ...). In that case, we have *destructive interference.* The crests line up with the troughs. The two waves cancel, so the sound is quiet.

1. Two speakers on the x axis emit the same 500 Hz tone. A person on the x axis thinks the sound is annoying. What should the minimum separation between the speakers be to make the sound quiet?

The speakers should be $\lambda/2$ apart for destructive interference.
The speed of sound in air is $v = 343$ m/s.

$v = \lambda f$, or $\lambda = v/f$
$\lambda = 343/500 = 0.686$ m, so $\lambda/2 =$ **0.343 m**

2. Two speakers emit identical 686 Hz sound waves. Does the person hear a quiet or loud sound?

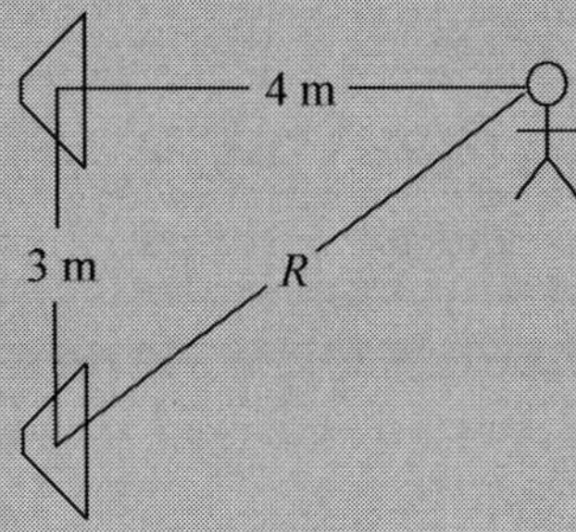

$\lambda = v/f = 343/686 = 0.5$ m

From Pythagoras, $R = \sqrt{3^2 + 4^2} = 5\,\text{m}$
The path-length difference is 5 m – 4 m = 1 m. This is 2λ.

The person is at a point of **constructive interference**, so the sound is **loud**.

Quiz 13.3

1. When strings from two instruments are played together, you can hear a beat frequency of 2 Hz. One instrument emits a 110 Hz tone. The second instrument is slightly sharp (higher frequency). What is its frequency?

2. In the previous problem, find the wavelengths of the sound waves from each instrument.

3. Two speakers and a listener are along a common line. The speakers emit identical sound waves of frequency 654 Hz. Because of their size, the speakers must be at least 1.2 m apart. What is the minimum distance between the speakers such that the listener will experience constructive interference?

4. A person listens to two speakers, which emit identical 257 Hz tones. Is the person in a loud or quiet spot?

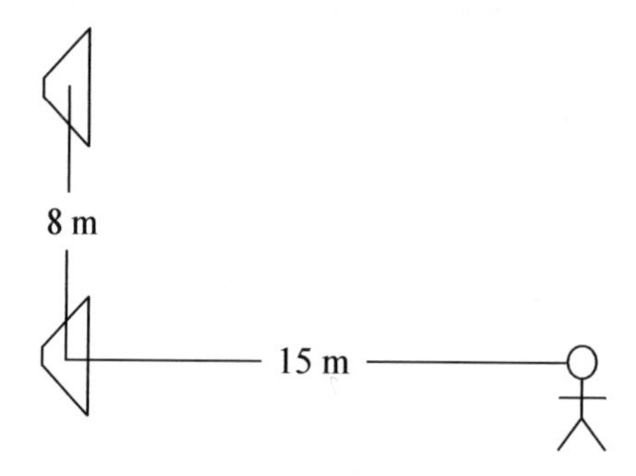

Chapter summary

Wavelength (λ) — Distance (m) from crest to crest

Frequency (f) — Waves passing by per second (s^{-1}, or Hz)

Wave speed (v) — $v = \lambda f$ (m/s)

Sinusoidal traveling wave — $y = A\cos(kx \pm \omega t)$
A = amplitude
$\lambda = 2\pi/k$
$f = \omega/(2\pi)$

Standing waves

- Clamped string: Node at each end
- Tube: Node at an open end, antinode at a closed end

Beat frequency — $f_{beat} = |f_1 - f_2|$

Interference

- Constructive: Path-length difference = 0, λ, 2λ...
- Destructive: Path-length difference = ½λ, 1½λ, 2½λ...

End-of-chapter questions

1. Light is a type of wave. The speed of light is 3×10^8 m/s. Orange light has a wavelength of 6×10^{-7} m. Find the frequency.

2. A snapshot of a wave at $t = 0$ is shown below. It travels to the left at 3 m/s. Write an equation for the wave.

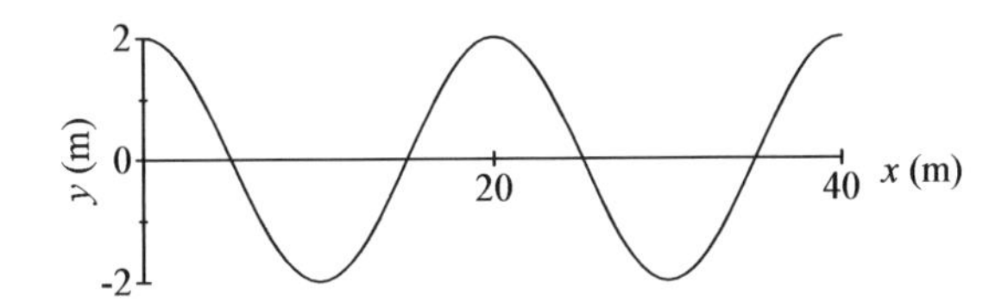

3. A wave on a string is given by $y = (7\text{ cm})\cos(4\pi x - 12\pi t)$, where x is in m and t is in s. Find the amplitude, wavelength, frequency, wave speed, and direction.

4. A 0.6 m long tube is open at both ends. The speed of sound is 343 m/s. What is the frequency of the fundamental mode?

5. A student determines that a traveling wave on a clamped, 5 m long string has a speed of 20 m/s. What are the frequencies of the first three standing-wave modes?

6. What is the beat frequency that occurs when a 440 Hz and 440.1 Hz tuning fork are played simultaneously? How many seconds are between each beat?

7. Two speakers are 50 m and 70 m away from a listener. The speakers play identical 68.6 Hz tones. Is the listener in a quiet or loud spot?

Challenging problems

1. A water wave on a pond has an amplitude 20 cm, speed 1 m/s, and wavelength 2 m. At $t = 0$, a duck sits in the trough of the wave. What is the duck's height, relative to equilibrium, at $t = 0.20$ s?

2. What are the wavelengths for these waves?

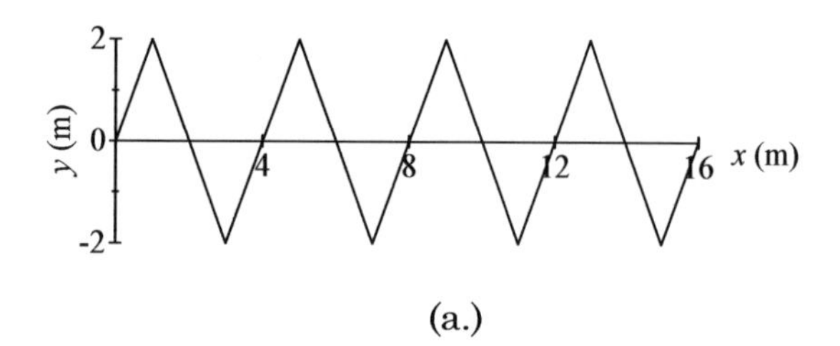

(a.)

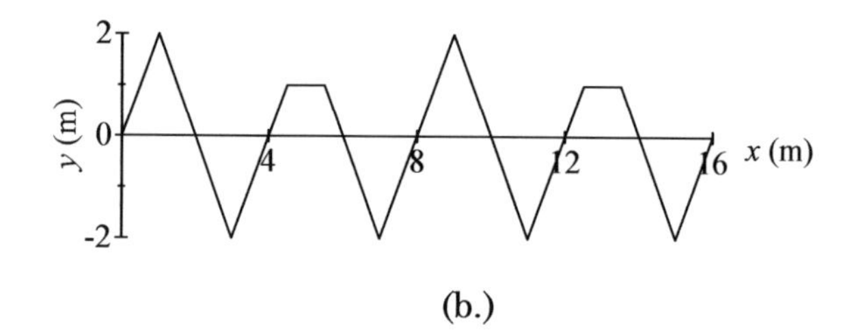

(b.)

3. Find the frequencies of the first two modes for a 0.5 m long tube that is closed at one end and open at the other end.

4. Water waves of wavelength 0.1 m travel through two small holes that are 0.5 m apart. A bug floats on the water. Will the bug's oscillations be large or small?

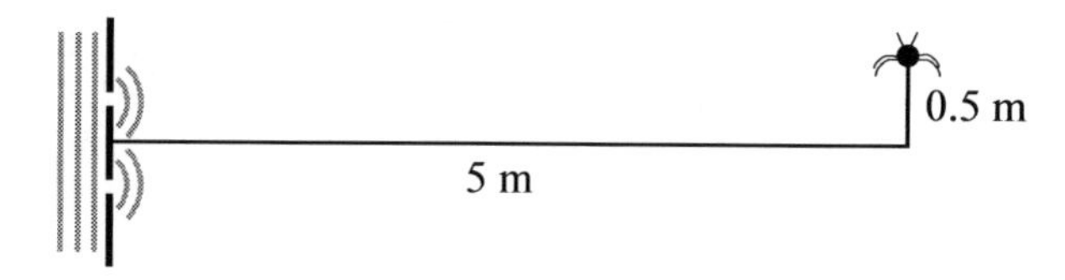

14 Fluids

Pressure

Pressure (P) is defined as force per area,

$$P = F/A$$

P is measured in units of N/m^2, or Pascals (Pa). There are many other units of pressure in common use, such as pounds per square inch (psi). One atmosphere of pressure (1 atm) is 101.3 kPa = 1.013×10^5 Pa.

Right now, gas molecules are colliding with your skin. These collisions produce 1 atm of pressure, which is quite a lot. Fortunately, your insides (lungs, etc.) are at an equal pressure, so you don't implode!

In a liquid such as water, the pressure increases with depth, due to the weight of the liquid above that depth. If the liquid is not moving, the pressure is given by

$$P = P_0 + \rho g d$$

where P_0 is the pressure at the surface, ρ is the density of the fluid (kg/m^3), and d is the depth (m). This equation does not apply to gases. The density of water is 1000 kg/m^3; i.e., a cubic meter has a mass of 1000 kg. (When calculating pressure, we'll use $g = 9.8$ m/s^2 rather than 10 m/s^2, for better accuracy.)

1. What is the force exerted by the air on one side of a 2 m × 2 m × 2 m cube?

$P = F/A$, or $F = PA$

The air pressure is $P = 1.013 \times 10^5$ Pa.
The area of one side is $A = 2 \times 2 = 4$ m^2.

$F = (1.013 \times 10^5)(4) =$ **4×10^5 N**

2. A scuba diver is at a depth of 20 m. What is the water pressure?

The pressure at the surface is 1 atm. $P_0 = 1.013 \times 10^5$ Pa.

$P = P_0 + \rho g d$
$= 1.013 \times 10^5 + (1000)(9.8)(20) =$ **3×10^5 Pa**, or **3 atm**

Static fluids

A *fluid* is defined as a substance that does not resist deformation. We can place it in a container of any shape. Examples include air, water, and oil. A *static fluid* is a fluid that is not moving.

A connected liquid is one where the liquid molecules are free to move from one region to another. In a connected static liquid, if we draw a horizontal line, the pressure is the same everywhere along the line.

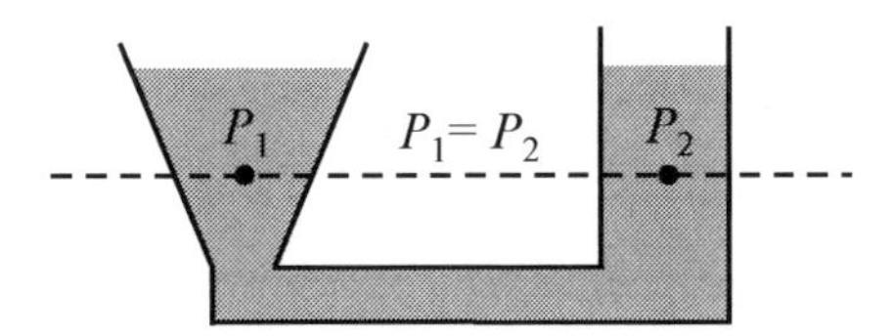

1. A U-tube contains liquid mercury ($\rho = 1.36 \times 10^4$ kg/m³). One side is a vacuum ($P = 0$) and the other side has a pressure 1 atm. Find the height difference h.

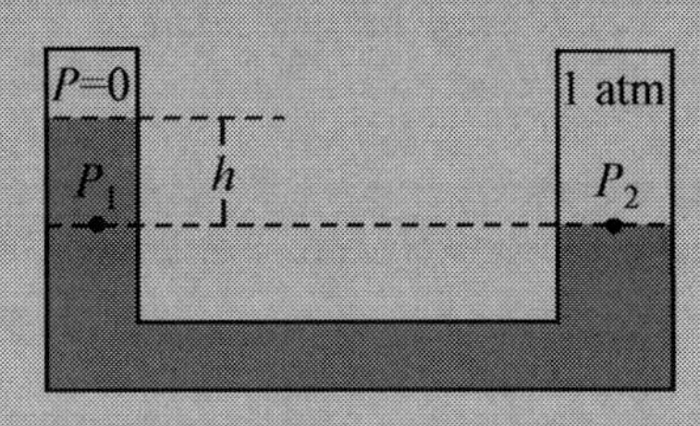

$P_1 = 0 + \rho gh = (1.36 \times 10^4)(9.8)h$
$P_2 = 1 \text{ atm} = 1.013 \times 10^5 \text{ Pa}$

Set them equal: $(1.333 \times 10^5)h = 1.013 \times 10^5$
$h = \underline{\mathbf{0.76\ m}}$ or $\underline{\mathbf{760\ mm}}$

1 atm pressure is sometimes called "760 mm Hg" or "760 Torr."

2. The column of water has a height $h = 0.15$ m. Find the pressure P.

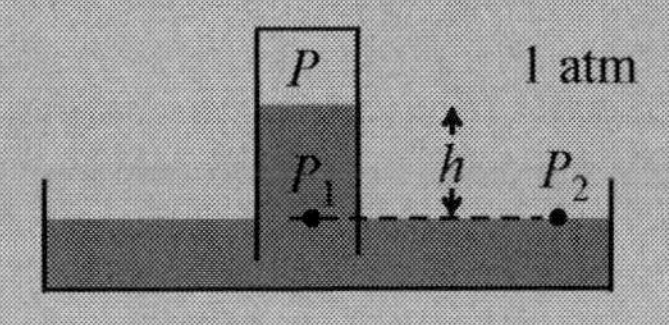

$P_1 = P + \rho gh = P + (1000)(9.8)(0.15) = P + 1470$
$P_2 = 1 \text{ atm} = 1.013 \times 10^5 \text{ Pa}$

Set them equal: $P + 1470 = 1.013 \times 10^5$
$P = \underline{\mathbf{9.98 \times 10^4\ Pa}}$

Quiz 14.1

1. A submarine is in the water. The water pressure is 7 atm. A hatch on the outside of the submarine has an area 2.3 m^2. How much force does the water exert on the hatch?

2. In the previous problem, what was the submarine's depth?

3. A U-tube contains water. One end is open to the air. The other end contains gas with a pressure 1.020×10^5 Pa. Find the height difference h.

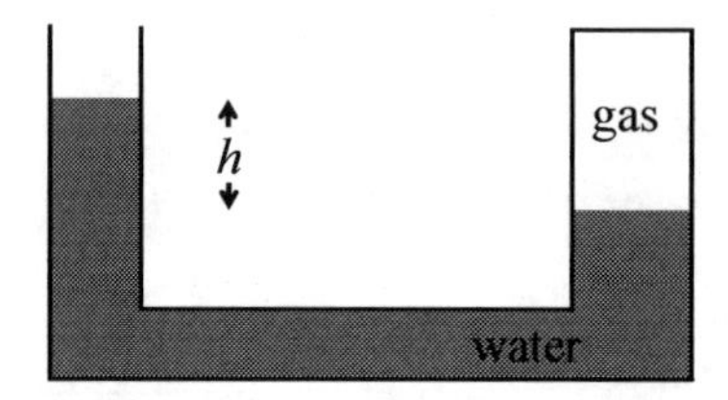

4. Suppose the air pressure is $P = 1.005 \times 10^5$ Pa. Find the height h of the mercury column, to the nearest mm.

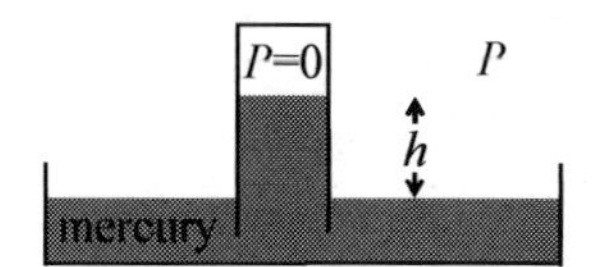

Buoyancy

In a liquid, pressure increases with depth. Suppose an object is submerged in the liquid. The bottom surface of the object experiences a larger pressure than the top. This pressure difference pushes the object up. The upward force is called *buoyancy*.

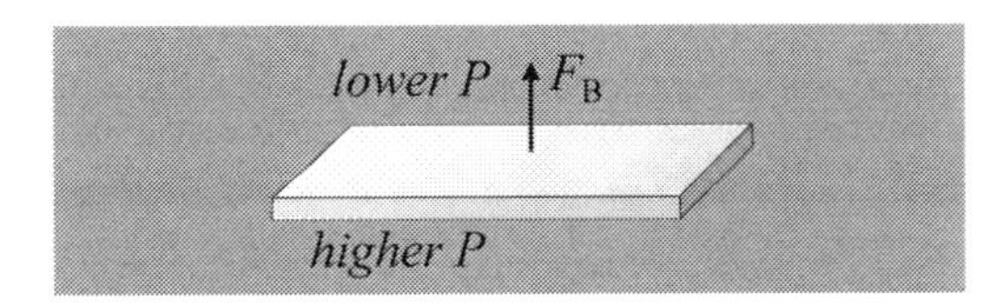

The magnitude of the buoyant force equals the weight of the displaced fluid,

$$F_B = \rho_f V_f g$$

where ρ_f is the fluid density (kg/m³) and V_f is the volume (m³) of the displaced fluid. The displaced fluid is the fluid that is pushed out of the way because of the object.

1. A rock of density 5000 kg/m³ and volume 0.01 m³ is completely submerged in water. Find the total force on the rock.

Since the rock is completely underwater, V_f equals the volume of the rock.

Buoyant force (up): $\rho_f V_f g = (1000)(0.01)(9.8) = \underline{98\text{ N}}$
The mass of the rock: $m = (5000\text{ kg/m}^3)(0.01\text{ m}^3) = 50\text{ kg}$
Gravity force (down): $mg = (50)(9.8) = \underline{490\text{ N}}$

The net force is 98 – 490 = **–392 N (down)**

2. A raft of mass 40 kg floats on the water. It is square with sides 2 m and height 0.25 m. Find x, the height that is under water.

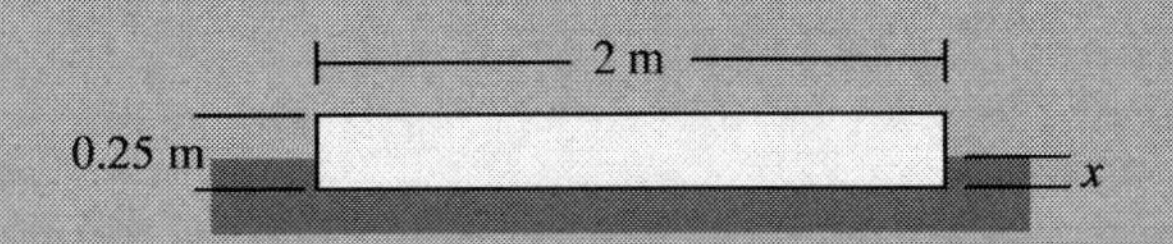

Volume of displaced fluid: $V_f = (2)(2)(x) = 4x$

Buoyant force (up): $\rho_f V_f g = (1000)(4x)(9.8) = 39{,}200x$
Gravity force (down): $mg = (40)(9.8) = 392$

Since the object floats, the buoyant force balances gravity.

$39{,}200x = 392$
$x = \mathbf{0.01\text{ m}}$

Moving fluids

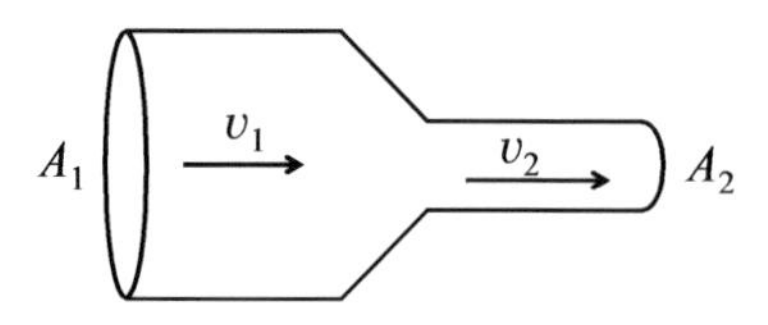

Suppose water is flowing through a pipe. The *volume flow rate* (Q) is the volume of water that flows through the pipe area every second. Q has units of m^3/s. For a fluid flowing at a speed v through a cross-sectional area A,

$$Q = Av$$

If the pipe narrows or widens, assuming steady flow, Q must remain constant. Each second, the same amount of water flows through the wide section and the narrow section. This gives us the *equation of continuity*,

$$A_1v_1 = A_2v_2$$

Since 1 m^3 is a large volume, sometimes liters (L) are used. 1 L = 10^{-3} m^3.

1. 7 liters of water flow through a pipe of radius 5 cm every second. Find the water speed.

$7\text{ L} = 7 \times 10^{-3}\text{ m}^3$, so $Q = \underline{7 \times 10^{-3}\text{ m}^3/\text{s}}$

$A = \pi r^2 = (3.14)(0.05^2) = \underline{7.85 \times 10^{-3}\text{ m}^2}$

$Q = Av$
$7 \times 10^{-3} = (7.85 \times 10^{-3})v$
$\underline{\mathbf{0.9\text{ m/s}}} = v$

2. Air flows through a 10 cm diameter tube at a speed 4 m/s. The tube narrows to a 2 cm diameter. What is the air speed in the narrow section?

$A_1 = \pi r^2 = (3.14)(0.05^2) = 7.85 \times 10^{-3}\text{ m}^2$

$A_2 = \pi r^2 = (3.14)(0.01^2) = 3.14 \times 10^{-4}\text{ m}^2$

$A_1v_1 = A_2v_2$
$(7.85 \times 10^{-3})(4) = (3.14 \times 10^{-4})v_2$
$\underline{\mathbf{100\text{ m/s}}} = v_2$

Quiz 14.2

1. A plastic toy has a density 950 kg/m^3 and volume 10^{-4} m^3. It is completely submerged under water. Find the total force on the toy.

2. A 0.6 kg wooden cube has dimensions 0.1 m × 0.1 m × 0.1 m. It floats on the water. What is the height of the cube under water?

3. 800 L/s of air flows through a 0.25 m diameter duct. Find the air speed.

4. A wine bottle has a 7.5 cm diameter at the base and 2.5 cm at the opening. Wine pours out the opening at 0.09 m/s. What is the speed of the wine flow at the base?

Bernoulli's equation

Suppose a fluid travels through a pipe. We trace the path of the fluid.

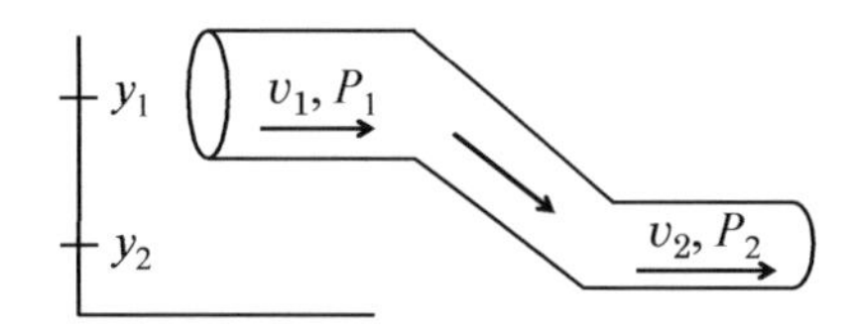

Bernoulli's equation relates the pressure (P) and speed (v) of the fluid at any two points:

$$P_1 + \tfrac{1}{2}\rho v_1^2 + \rho g y_1 = P_2 + \tfrac{1}{2}\rho v_2^2 + \rho g y_2$$

where y is the height. Note that we ignore viscosity (resistance to flow).

1. Water flows through a wide horizontal tube at a speed $v_1 = 4$ m/s. The tube narrows and the water shoots into the air with a speed $v_2 = 16$ m/s. What is the water pressure in the wide section?

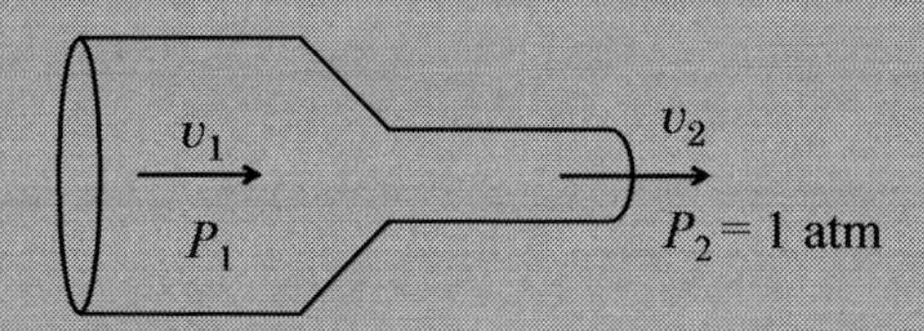

$P_1 + \tfrac{1}{2}\rho v_1^2 + \rho g y_1 = P_2 + \tfrac{1}{2}\rho v_2^2 + \rho g y_2$
Since the tube is horizontal, we can set $y_2 = y_1$ and drop from the equation.

$P_1 + \tfrac{1}{2}(1000)(4^2) = 1.013 \times 10^5 + \tfrac{1}{2}(1000)(16^2)$
$P_1 =$ $\underline{\mathbf{2.213 \times 10^5\ Pa}}$

2. Water is pumped through a pipe up a 10 m high hill. At the bottom of the hill, water flows 2 m/s and a pressure gauge reads 200.5 kPa. At the top, the pressure gauge reads 100.0 kPa. What is the water speed at the top?

$P_1 + \tfrac{1}{2}\rho v_1^2 + \rho g y_1 = P_2 + \tfrac{1}{2}\rho v_2^2 + \rho g y_2$

$2.005 \times 10^5 + \tfrac{1}{2}(1000)(2^2) + 0 = 1.000 \times 10^5 + \tfrac{1}{2}(1000)v_2^2 + (1000)(9.8)(10)$
$1.005 \times 10^5 - 96{,}000 = (500)v_2^2$

$9 = v_2^2$, so $v_2 =$ $\underline{\mathbf{3\ m/s}}$

Streamlines

Imagine we painted three air molecules black and monitored their paths as they flowed through a tube. They would trace lines called *streamlines.*

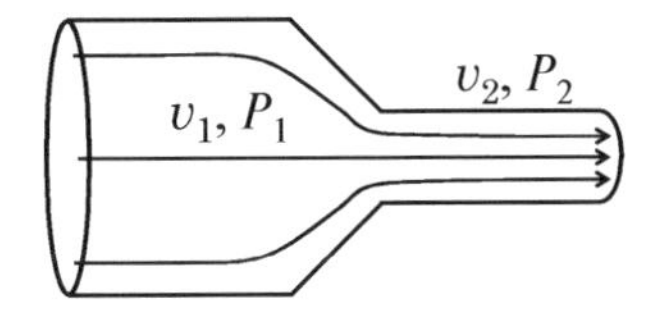

In the picture, $v_1 < v_2$. The streamlines are spaced far apart in the wide section, where the velocity is low. From Bernoulli's equation, if velocity is low, then pressure is high. Therefore,

- If the streamlines are far apart, the velocity is low and the pressure is high.
- If the streamlines are close together, the velocity is high and the pressure is low.

1. Wind blows over the roof of a house. Is the pressure above the roof higher or lower than the pressure inside the house?

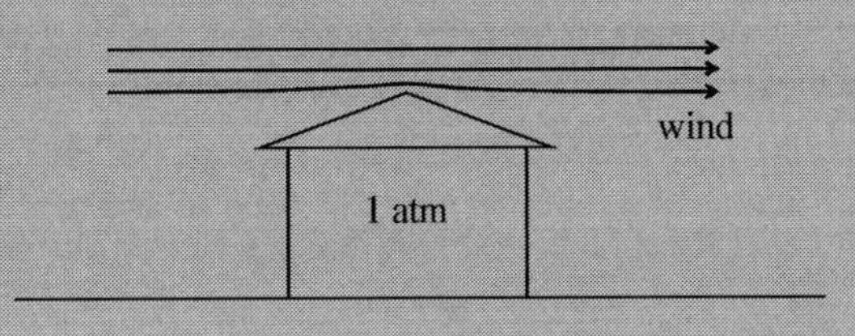

The high air speed of the wind means that the pressure is **lower** than in the house. The pressure difference causes an upward force, which might tear off the roof!

2. Air (density 1.2 kg/m³) flows through a horizontal tube of cross-sectional area 0.1 m² at a speed 50 m/s and pressure 102.0 kPa. The tube narrows to 0.05 m². What is the pressure in the narrow section?

$A_1v_1 = A_2v_2$
$(0.1)(50) = (0.05)v_2$
$\underline{100\text{ m/s}} = v_2$

$P_1 + \frac{1}{2}\rho v_1^2 + \rho g y_1 = P_2 + \frac{1}{2}\rho v_2^2 + \rho g y_2$
Since the tube is horizontal, we can set $y_2 = y_1$ and drop from the equation.

$1.020 \times 10^5 + \frac{1}{2}(1.2)(50^2) = P_2 + \frac{1}{2}(1.2)(100^2)$
$\underline{\mathbf{9.75 \times 10^4\ Pa}} = P_2$

Quiz 14.3

1. Water flows at 5 m/s through a narrow horizontal pipe. A pump maintains a 102 kPa pressure. The pipe widens. In the wide section, the water speed is 1 m/s. What is the pressure in the wide section?

2. Oil (density 900 kg/m^3) is poured into a pipe at the top of a 20 m high hill, where the pressure is 1 atm. The oil starts from rest. At the bottom of the hill, the oil flows at 10 m/s. What is the oil pressure at the bottom?

3. Air flows over a plastic ball. What direction is the force on the ball (up or down)?

4. Water flows through a horizontal, narrow tube of area 0.03 m^2, at a speed 5 m/s. The tube widens to 3.0 m^2. A pressure gauge in the wide section reads 123 kPa. What is the pressure reading in the narrow section?

Chapter summary

Pressure (N/m², or Pa)	$P = F/A$
Static fluids	
Pressure in a static liquid	$P = P_0 + \rho g d$
Buoyant force	$F_B = \rho_f V g$
Moving fluids	
Volume flow rate (m³/s)	$Q = Av$
Equation of continuity	$A_1 v_1 = A_2 v_2$
Bernoulli's equation	$P_1 + \frac{1}{2}\rho v_1^2 + \rho g y_1 = P_2 + \frac{1}{2}\rho v_2^2 + \rho g y_2$
Atmospheric pressure (1 atm)	1.013×10^5 Pa
Density of water	1000 kg/m³

End-of-chapter questions

1. A scuba diver is at a depth of 20 m. His mass is 80 kg.

   a. What is the water pressure? Express your answer to the nearest atm.
   b. Suppose the buoyant force exactly balances the diver's weight. What is the volume of the diver?

2. A 10 kg mass floats on the surface of oil (density 900 kg/m³). The oil is in a tall, 4 cm diameter cylinder.

   a. What is the pressure at the surface of the oil?
   b. What is the pressure at a depth of 100 cm?

3. A bucket is filled with water. There is a small hole 0.2 m below the surface. What is the speed of the water shooting out of the hole? (The water velocity at point 1 is nearly zero.)

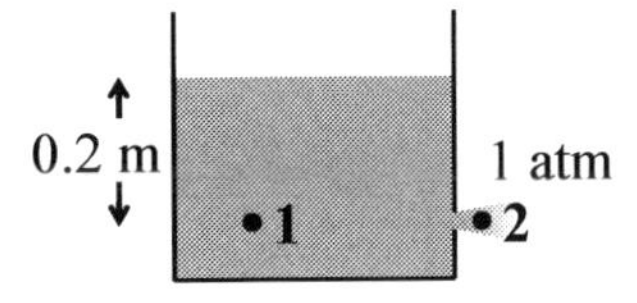

4. Water flows through garden hose at 10 m/s. A gardener puts a thumb over the end, reducing the cross-sectional area by 50%.

   a. What is the speed of the water that exits the hose?
   b. What is the pressure of the water in the hose?

5. A sprinkler consists of a 1 m tall vertical pipe with a head on top. The head has many small holes to let the water out. At the bottom of the pipe, the water pressure is 2.5 atm and the water flows up at 2 m/s. What is the water speed as it exits the head?

Challenging problems

1. A 30 kg ball of diameter 0.4 m hangs from a rope. The ball is halfway submerged in water. What is the tension of the rope?

2. "Gauge pressure" is the pressure relative to 1 atm (760 mm Hg). Blood pressure, for instance, is measured this way. A blood pressure reading of 120 mm Hg means that the actual pressure is 120 + 760 = 980 mm Hg. The blood pressure of a giraffe is 280 mm Hg near its heart. If its head is 2.5 m above the heart, what is its blood pressure there? Assume a blood density of 1000 kg/m^3.

3. Air (density 1.2 kg/m^3) flows through a wide tube of cross-sectional area 0.04 m^2 at a speed 10 m/s. The tube narrows to 0.01 m^2. A U-tube, which contains mercury, connects the wide and narrow sections. Find the height difference h.

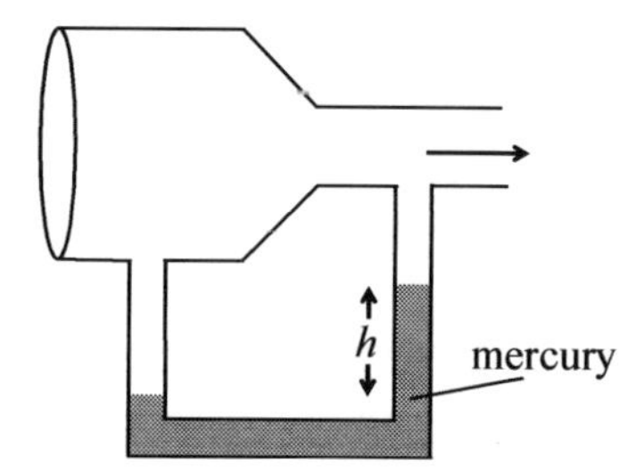

4. Water flows out of a faucet at 6 L/min. The opening of the faucet has a diameter of 1 cm. The water flows vertically downward.

   a. What is the water speed 10 cm below the faucet opening?
   b. What is the diameter of the water column 10 cm below the faucet opening?

15 Thermodynamics

Ideal gas law

Most gases obey the ideal gas law,

$$PV = nRT$$

where P is pressure (Pa), V is volume (m^3), n is the number of moles (mol), $R = 8.314$ J/(mol·K) is the ideal gas constant, and T is the absolute temperature in Kelvin (K).

The temperature in K is the temperature in °C plus 273. For example, 25°C is 25 + 273 = 298 K.

There are 6.02×10^{23} molecules in 1 mole.

One mole has a mass (in grams) equal to the molecular mass. For example, an oxygen atom (O) has an atomic mass of 16. An oxygen molecule (O_2) has a molecular mass 16 + 16 = 32. One mole of oxygen gas is therefore 32 g.

1. Hydrogen gas is in a 0.5 m^3 container. It is held at 27°C and 1 atm. How many moles of gas are there?

$PV = nRT$
$(1.013 \times 10^5)(0.5) = n\ (8.314)(300)$
$\underline{\textbf{20 mol}} = n$

2. In the previous problem, what is the mass of the hydrogen gas?

A hydrogen molecule (H_2) has a molecular mass 1 + 1 = 2, so there are 2 g/mol.

$$\left(\frac{2\text{ g}}{\text{mol}}\right)(20\text{ mol}) = \underline{\textbf{40 g}}$$

3. 1 mole of oxygen is in a bottle at 2 atm and 0°C. What is the volume of the bottle in liters?

$PV = nRT$
$(2 \times 1.013 \times 10^5)V = (1)(8.314)(273)$
$V = 0.0112\ m^3$

$$\left(\frac{1\text{ L}}{10^{-3}\text{ m}^3}\right)(0.0112\text{ m}^3) = \underline{\textbf{11.2 L}}$$

Thermal energy

The pressure of a gas results from molecules colliding with the container walls. These collisions exert a force on the walls.

When temperature increases, the molecules move faster and have more kinetic energy. The *thermal energy* of an ideal gas is

$$E = \frac{3}{2} nRT$$

This is the total kinetic energy due to the motion of all the molecules in the gas. (We neglect rotational and vibrational kinetic energy).

Isothermal means the temperature is constant. *Isobaric* means the pressure is constant.

1. 4 moles of gas are heated from 300 to 400 K. What is the change in thermal energy?

E(before) = 3/2 (4)(8.314)(300) = 14,965 J
E(after) = 3/2 (4)(8.314)(400) = 19,954 J

ΔE = 19,954 – 14,965 = **4989 J**

2. A gas has a pressure 10^5 Pa and volume 10 m^3. It expands isothermally to 20 m^3. Find the pressure and thermal energy after expansion.

$PV = nRT$

Before: $(10^5)(10) = nRT$
$10^6 = nRT$

We assume the container does not leak, so n does not change.
Isothermal means T does not change, so nRT is constant.

After: $P(20) = 10^6$
P = **5×10^4 Pa**

$E = 3/2\ nRT$ = **1.5×10^6 J**

(For an isothermal process, since T does not change, E does not change.)

Quiz 15.1

1. Oxygen gas is in a 2 L bottle at 20°C and 1 atm. What is the mass of the oxygen gas?

2. 1 mole of hydrogen gas is in a container at 1 atm and 0°C. What is the volume of the container in liters?

3. 64 grams of oxygen gas are heated from 0°C to 25°C. What is the change in thermal energy?

4. A gas in a 10 L container has a pressure of 0.3 atm. It undergoes isothermal compression to 1.0 L. Find the pressure and thermal energy after compression.

Work and heat

There are two ways to change the thermal energy of a gas: work or heat. Recall from Chapter 7 that work is force times distance. Suppose you exert a force on a piston and compress a gas. You are doing positive work on the gas.

Assuming constant pressure, the work done on a gas is

$$W = -P\,\Delta V$$

where ΔV is the volume change. If the gas is compressed, $\Delta V < 0$ and $W > 0$. Positive work is being done *on* the gas. If the gas expands, $\Delta V > 0$ and $W < 0$. This means work is being done *by* the gas; e.g., an expanding piston that drives a crankshaft.

Heat (Q) is energy transferred from a high-temperature substance to a low-temperature one. For example, a flame transfers heat to a pan of water.

1. A gas at 2 atm undergoes isobaric expansion. It expands from 3 L to 10 L. How much work was done on the gas?

$$W = -P\,\Delta V$$
$$= -(2 \times 1.013 \times 10^5 \text{ Pa})(7 \times 10^{-3} \text{ m}^3)$$
$$= \underline{\mathbf{-1400\ J}}$$

(The work done *by* the gas is $P\,\Delta V = 1400$ J)

2. 3 moles of gas are in a metal container. The volume of the container is fixed. A professor holds a torch to the container and transfers 300 J of heat to the gas. What is the temperature change of the gas?

Because $\Delta V = 0$, no work is done on the gas. The change in energy is due to heat.

$$\Delta E = Q = 300 \text{ J}$$

$$\Delta E = \frac{3}{2}nR\Delta T$$

$$300 = (1.5)(3)(8.314)\Delta T$$

$$\underline{\mathbf{8\ K}} = \Delta T$$

Energy conservation

The thermal energy of a substance increases when work is done on it (positive W) or when heat is transferred into it (positive Q). The thermal energy decreases when it does work on the outside world (negative W) or when heat flows out (negative Q).

In general, the change in thermal energy is given by the *First Law of Thermodynamics*,

$$\Delta E = W + Q$$

For an ideal gas,

$$\Delta E = \frac{3}{2} nR\Delta T$$

1. A machine did 10 J of work compressing a gas. The compression was isothermal. What was the heat flow?

Isothermal means $\Delta T = 0$, so $\Delta E = 0$.

$\Delta E = W + Q$
$0 = 10 + Q$
$\underline{-10 \text{ J}} = Q$ (heat flowed out of the gas)

2. Helium gas is in a 10 L container. A wall of the container is free to slide. Ice cubes are placed on the container and the volume decreases to 8 L. The pressure, 1 atm, is constant. What is the heat flow?

$W = -P\Delta V = -(1.013 \times 10^5)(-2 \times 10^{-3}) = \underline{203 \text{ J}}$

$PV = nRT$

Before, $nRT = (1.013 \times 10^5)(10 \times 10^{-3}) = 1013$
After, $nRT = (1.013 \times 10^5)(8 \times 10^{-3}) = 810$

$$\Delta E = \frac{3}{2}(810) - \frac{3}{2}(1013) = \underline{-305 \text{ J}}$$

$\Delta E = W + Q$
$-305 = 203 + Q$
$\underline{-508 \text{ J}} = Q$ (heat flows out of the gas)

Quiz 15.2

1. A constant pressure of 1.3 atm is applied to a gas, causing it to compress from 2 L to 1 L. How much work was done on the gas?

2. 2 moles of gas are in a container with a fixed volume. Ice cubes are placed on the box, causing the temperature to decrease by 5°C. What was the heat flow?

3. A gas expands isothermally. It does 8 J of work. What is the heat flow?

4. An ideal gas expands from 3 m^3 to 4 m^3 at a constant pressure of 0.8 atm. What is the heat flow?

Heat engines

An engine is quite complicated, but here we just look at the basics. Gas is in a cylinder. A heat source such as a flame causes the gas to expand. When it expands, it pushes on a piston, which might turn a crankshaft. Then, the gas is placed in contact with a cold reservoir (e.g., air or water) and it contracts. This expansion-contraction cycle repeats over and over, effectively converting heat into useful work.

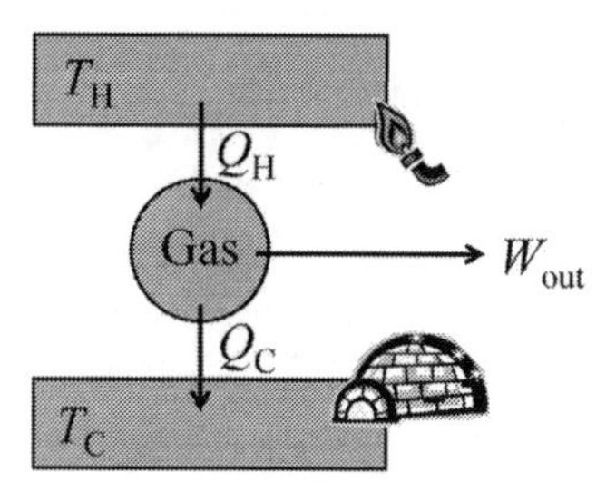

During the first part of the cycle, heat Q_H flows from the hot reservoir into the gas. During the second part, Q_C flows from the gas to the cold reservoir. Over the entire cycle the gas does work W_{out}. For an ideal engine, the work equals the heat that went in minus the heat that went out:

$$W_{out} = Q_H - Q_C$$

The *efficiency* (e) of an engine equals the work divided by how much heat we had to put in:

$$e = \frac{W_{out}}{Q_H}$$

The theoretical maximum efficiency is

$$e_{max} = 1 - T_C/T_H$$

where T_H and T_C are the temperatures of the hot and cold reservoirs, measured in K.

1. In a particular heat-engine cycle, 5 J of heat go into the gas and 3 J are exhausted. How much work is done in a cycle, and what is the efficiency?

$W_{out} = Q_H - Q_C = 5 - 3 =$ **2 J**

$e = 2/5 =$ **0.4** (or 40%)

2. What is the maximum efficiency of a heat engine that uses a 900°C flame as the hot reservoir and 25°C water as the cold reservoir?

$T_H = 900 + 273 = 1173$ K
$T_C = 25 + 273 = 298$ K

$e_{max} = 1 - T_C/T_H = 1 - 298/1173 =$ **0.75** (or 75%)

Entropy

If enough heat flows into an ice cube, it will melt, transforming from a solid to a liquid. In a solid, the molecules are fixed in specific locations. In a liquid, the molecules are free to move around. The liquid is more *disordered* than the solid. *Entropy* (S) is a measure of disorder. In this example, $S_{liquid} > S_{solid}$.

In general, if heat Q is put into a substance of temperature T (Kelvin), its entropy will increase by

$$\Delta S = Q/T$$

In this equation, we assume that T is constant while Q is being transferred.

The *Second Law of Thermodynamics* says that, for an isolated system, the total entropy will increase or stay the same: $\Delta S \geq 0$. It turns out that this is what limits the efficiency of heat engines, as defined in the previous section. The law also requires heat to spontaneously flow from a hot object to a cold object and not the other way around.

1. A hot object (375 K) is in contact with a cold object (300 K). 5 J of heat flows from hot to cold. What is the change in entropy of each object? What is the change in entropy of the system?

ΔS(hot object) = –5/375 = **–0.0133 J/K**

ΔS(cold object) = 5/300 = **0.0167 J/K**

ΔS(system) = –0.0133 + 0.0167 = **0.0034 J/K**

2. A gas at 300 K expands isothermally and does 50 J of work. What is the change in the entropy of the gas?

$\Delta E = W + Q$

$W = -50$ J (– because the gas is doing work on the outside world)
Isothermal: $\Delta E = 0$

$0 = -50 + Q$, or $Q = 50$ J

$\Delta S = 50/300 =$ **0.167 J/K**

Quiz 15.3

1. An engine does 100 J of work every second (100 W). Heat goes out the exhaust tailpipe at 80 J per second. What is the efficiency?

2. A hobbyist makes a heat engine, using boiling water (100°C) as the hot reservoir and ice (0°C) as the cold reservoir. What is the maximum possible efficiency?

3. A person with a body temperature of 37°C touches a piece of metal at 23°C. If 0.1 J of heat flows from the person to the metal, find the entropy change of the system (person + metal).

4. A piston does 8 J of work on a gas, causing it to compress. The temperature is held fixed, 400 K. What is the change in the entropy of the gas?

Chapter summary

Ideal gas law	$PV = nRT$ $R = 8.314$ J/(mol·K)
Temperature (K) = temperature (°C) + 273	
Thermal energy for an ideal gas	$E = \frac{3}{2}nRT$
Change in thermal energy	$\Delta E = \frac{3}{2}nR\Delta T$
Work done on a gas	$W = -P\Delta V$
First Law of Thermodynamics	$\Delta E = W + Q$
Heat engines	
Work done	$W_{out} = Q_H - Q_C$
Efficiency	$e = W_{out}/Q_H$
Maximum theoretical efficiency	$e_{max} = 1 - T_C/T_H$
Change in entropy	$\Delta S = Q/T$

End-of-chapter questions

1. A gas container is filled with 10 moles of helium atoms. The pressure is 1 atm and the temperature is 300 K.

    a. What is the volume of the container?
    b. Find the thermal energy of the gas.

2. In the previous problem, someone puts a flame to the container, raising the gas temperature to 400 K. The pressure remains constant.

    a. What is the change in volume?
    b. What is the change in thermal energy?
    c. How much heat went into the gas?

3. 2 moles of gas are in a metal box with a fixed volume. Cold water is spilled on the box, causing the gas temperature to fall from 298 to 288 K. Find the heat flow.

4. A heat engine has a 200°C hot reservoir at and a 40°C cold reservoir.

    a. What is the maximum possible efficiency?
    b. Assuming this efficiency, how much work can the engine do, if 50 J of heat is supplied by the hot reservoir?

5. Water in a puddle has a temperature of 0°C. It slowly freezes, releasing 10 J of heat to the air. The air temperature is -20°C. What is the total change in entropy?

Challenging problems

1. A gas has an initial volume 12 L, temperature 300 K, and pressure 1 atm. The container's top surface is a piston (area 0.01 m^2) that can move up or down. Someone places a 20 kg mass on the piston, which compresses the gas. The temperature stays fixed.

   a. What is the volume of the compressed gas?
   b. Did the entropy of the gas increase or decrease?

2. 4 moles of gas are in an insulating container, which does not allow heat to flow in or out. The initial temperature is 300 K. The container expands, doing 1000 J of work on the outside world. What is the temperature of the expanded gas? (This is called *adiabatic* expansion).

3. At 1 atm and 25°C, a diver inhales 0.25 moles of air with each breath. The diver goes into the water, to a depth of 30 m. The temperature of the scuba air is 20°C. How many moles are inhaled with each breath?

4. A refrigerator is an example of a *heat pump*. A heat pump is like a heat engine, except that work is done *on* the gas. Q_C is taken from the cold reservoir (e.g., the inside of the refrigerator) and Q_H is put into the hot reservoir (e.g., the kitchen). Suppose a heat pump does 6 J of work and puts 14 J of heat into the hot reservoir. How much heat was taken from the cold reservoir?

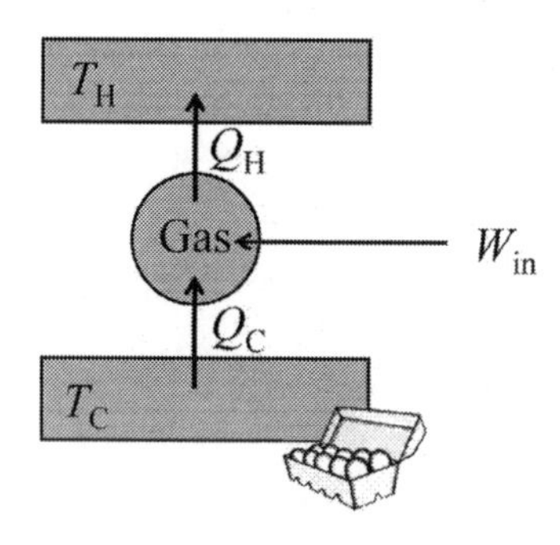

Answers

Chapter 1

Quiz 1.1 1.) displacement = 45 m, velocity = 5 m/s; 2.) displacement = -40 cm, velocity = -8 cm/s; 3.) 50 m/s^2; 4.) -10 m/s^2

Quiz 1.2 1.) 0 m/s; 2.) 30 m/s; 3.) 19 m; 4.) -5 m

Quiz 1.3 1.) 2 s; 2.) 10 s; 3.) 3 s; 4.) 20 s

End-of-chapter questions

1.) displacement = -27 mm, velocity = -0.5 mm/s; 2.) -4 m/s^2; 3.) 25 m/s; 4.) 500 m; 5.) 570 m; 6.) 10 s; 7.) 5 s; 8.) 7 s

Challenging problems

1.) (a) 1 s, (b) 4 m/s; 2.) (a) 5 s, (b) 300 m; 3.) (a) 1 s, (b) -10 m/s; 4.) 0.7 s

Chapter 2

Quiz 2.1 1.) v_x = 2 m/s, v_y = -4 m/s; 2.) 13 m/s ; 3.) 60°; 4.) 44 m/s

Quiz 2.2 1.) (7 m, -47 m); 2.) 4 s; 3.) v_x = 34 m/s, v_y = -40 m/s

Quiz 2.3 1.) 2 s; 2.) 20 m ; 3.) 2 s; 4.) 8 m

End-of-chapter questions

1.) v_x = 71 m/s, v_y = 71 m/s; 2.) 3 s; 3.) v_x = 7 m/s, v_y = -30 m/s; 4.) 4 s; 5.) 80 m; 6.) 0.5 s; 7.) 2.75 m

Challenging problems

1.) 2 s; 2.) 1 s; 3.) (a) 5 s, (b) 43 m; 4.) 3.7 m

Chapter 3

Quiz 3.1 1.) 0.8 m/s^2; 2.) a_x = -0.07 m/s^2, a_y = -0.14 m/s^2; 3.) F_y = -80 N, weight = 80 N; 4.) a_y = -10 m/s^2 (both)

Quiz 3.2 1.) F_x = 5 N; 2.) F_y = -10 N; 3.) 3000 N; 4.) 0.05 N

Quiz 3.3 1.) 10 N; 2.) 400 N; 3.) 300 N; 4.) 2.5 m/s^2

End-of-chapter questions

1.) a_x = 0.5 m/s^2, a_y = -0.25 m/s^2; 2.) F_x = -10,300 N; 3.) 10 N; 4.) 5 N; 5.) 15 N; 6.) 1500 N; 7.) 3.7 m/s^2

Challenging problems

1.) (a) 1000 N, (b) 2000 N; 2.) 134 N; 3.) 6 s; 4.) (a) 1 N, (b) 3 m/s^2

Chapter 4

Quiz 4.1 1.) $F = 8.5$ N, $\theta = 45°$ below horizontal; 2.) 10.4 N; 3.) $a_x = 1$ m/s^2, $a_y = 1$ m/s^2; 4.) $a_x = -1.23$ m/s^2, $a_y = 2$ m/s^2

Quiz 4.2 1.) 29 N; 2.) 2 m/s^2; 3.) 22.3 N; 4.) 0.9 N (+x direction)

Quiz 4.3 1.) $a_x = 7$ m/s^2 ; 2.) 28.2 N; 3.) 21.8°; 4.) $a_x = 7.7$ m/s^2

End-of-chapter questions

1.) $a_x = -0.62$ m/s^2, $a_y = 0.75$ m/s^2; 2.) 20 N; 3.) $a_x = 0.19$ m/s^2; 4.) 977 N; 5.) $a_x = 5$ m/s^2; 6.) $a_x = 2.8$ m/s^2; 7.) 0.5

Challenging problems

1.) (a) 480 N, (b) 12 m; 2.) (a) $a_x = -5$ m/s^2, (b) $a_x = -10$ m/s^2, $a_y = 7.3$ m/s^2; 3.) (a) 78.8 N, (b) 13.9 N; 4.) 1193 N

Chapter 5

Quiz 5.1 1.) $a_y = -8$ m/s^2; 2.) 30 N; 3.) 26 m/s^2; 4.) 250 N

Quiz 5.2 1.) $a_y = -6$ m/s^2; 2.) 160 N; 3.) $a_x = -3$ m/s^2; 4.) 21 N

Quiz 5.3 1.) 25 kg; 2.) 50 N; 3.) $a_y = 2$ m/s^2; 4.) 60 N

End-of-chapter questions

1.) (a) 50 N, (b) 25 N; 2.) (a) 50 N, (b) $a_y = 3.3$ m/s^2; 3.) $a_y = -2.5$ m/s^2; 4.) (a) 50 N, (b) 6 kg; 5.) $a_y = -2.35$ m/s^2

Challenging problems

1.) $a_y = 3$ m/s^2; 2.) 3 s; 3.) $a_x = -1.75$ m/s^2; 4.) $a_x = 1.3$ m/s^2

Chapter 6

Quiz 6.1 1.) 1 m/s^2, toward center; 2.) 5900 m/s^2, toward center; 3.) 128 N; 4.) 10 N

Quiz 6.2 1.) 50 N; 2.) 10 m/s; 3.) 0.25 N; 4.) 0.4

Quiz 6.3 1.) 18 N; 2.) 78 N; 3.) 420 N; 4.) 15.8 m/s

End-of-chapter questions

1.) 6×10^{-3} m/s^2, toward sun; 2.) 1.7×10^{-3} N; 3.) 0.016 N; 4.) 35 m/s; 5.) (a) 31 N, (b) 111 N; 6.) 3 m/s

Challenging problems

1.) 71 m/s; 2.) (a) 8 N, (b) 0.32 N; 3.) 71 N; 4.) (a) 0.15 N, (b) 0.14 s

Chapter 7

Quiz 7.1 1.) 14 J; 2.) 250 J; 3.) -250 J; 4.) 5×10^5 J

Quiz 7.2 1.) 8 m/s; 2.) -40 J; 3.) 20 m/s; 4.) 10 m/s

Quiz 7.3 1.) 1260 J; 2.) -26 J; 3.) 5.7 m/s; 4.) 1500 W

End-of-chapter questions

1.) 1.2×10^5 m/s; 2.) -15 J; 3.) 6.3 m/s; 4.) (a) 5 m/s, (b) 15 m/s; 5.) 6×10^5 J; 6.) (a) -30 J, (b) 7.6 m/s; 7.) 1400 W

Challenging problems

1.) -30 J; 2.) (a) 46 m/s, (b) 103 m; 3.) 152 W; 4.) 450 J

Chapter 8

Quiz 8.1 1.) 10^5 kg m/s; 2.) 20,000 N; 3.) 0.5 m/s; 4.) 23.75 J

Quiz 8.2 1.) 1 m/s, $-x$ direction; 2.) 27 J; 3.) $v_x = 5$ m/s, $v_y = 3$ m/s; 4.) $v = 5.8$ m/s, $\theta = 31°$

Quiz 8.3 1.) 0.2 m; 2.) 0.02 m/s; 3.) -0.1 m/s

End-of-chapter questions

1.) 300 N; 2.) (a) 1 m/s, (b) 12 J; 3.) 10 m/s; 4.) $v = 8.7 \times 10^3$ m/s, $\theta = 41°$ below horizontal; 5.) 0.032 m; 6.) 2×10^{-8} m/s

Challenging problems

1.) $v_1 = v_2 = 7.1$ m/s; 2.) 0.0125 m; 3.) 0.87 m/s; 4.) (a) -3 m/s, (b) 54 J

Chapter 9

Quiz 9.1 1.) $\omega = 7.3 \times 10^{-5}$ rad/s; 2.) 467 m/s; 3.) 2.6×10^{29} J; 4.) 0.009 J

Quiz 9.2 1.) 0.15 kg·m²; 2.) 12 kg·m²; 3.) 4 N·m; 4.) 1.2 N·m

Quiz 9.3 1.) -1.3 N·m; 2.) 61.4 N·m; 3.) 0.021 N·m; 4.) -0.5 rad/s²

End-of-chapter questions

1.) (a) 1.6 rad/s, (b) 3.1 m/s, tangential; 2.) 2.6×10^5 J; 3.) 256 J; 4.) 2300 N·m; 5.) 0.89 rad/s²

Challenging problems

1.) (a) -90 N·m, (b) 12 J; 2.) 17 N·m; 3.) (a) 16 J, (b) 2 rad/s; 4.) (a) 600 kg·m², (b) 740 J

Chapter 10

Quiz 10.1 1.) 7.1×10^{33} kg·m²/s; 2.) -7.5×10^{6} kg·m²/s; 3.) 24 rad/s; 4.) 3.75 rad/s

Quiz 10.2 1.) 0.75 rad/s; 2.) 0.067 rad/s; 3.) 0.75 rad/s

Quiz 10.3 1.) 2 m/s; 2.) 0.13 m; 3.) 1.6 rad/s; 4.) 40 rpm

End-of-chapter questions

1.) 0.04 kg·m²/s; 2.) 2.2 rad/s; 3.) 30 rpm; 4.) 3 rad/s; 5.) 4 rad/s, counterclockwise

Challenging problems

1.) 51 km/s; 2.) 80 rpm, clockwise; 3.) 0.06 rad/s; 4.) (a) 6.5 m/s, (b) 110 N

Chapter 11

Quiz 11.1 1.) 7.5 m; 2.) $x_{cg} = 0.3$ m, $y_{cg} = 0.9$ m; 3.) 4430 km; 4.) $x_{cg} = -0.167$ m, $y_{cg} = 0.167$ m

Quiz 11.2 1.) $F_1 = 242$ N, $F_2 = 208$ N; 2.) 23 N; 3.) 900 N, down; 4.) 313 N

Quiz 11.3 1.) 1.2 m; 2.) 0.7 m; 3.) 53°; 4.) 0.85 m

End-of-chapter questions

1.) (a) 1.07 m, (b) 2.7°; 2.) Kevin = 95 N, Rodney = 65 N; 3.) 48 N; 4.) 1.5 m

Challenging problems

1.) (a) 40 N, (b) $F_x = 35$ N, $F_y = 20$ N; 2.) 17 N; 3.) 40 N; 4.) $x = 0.35$ m

Chapter 12

Quiz 12.1 1.) 5.1 N; 2.) 200 N/m; 3.) amplitude = 7 cm, frequency = 5 Hz, period = 0.2 s; 4.) 0.5 s

Quiz 12.2 1.) 25 cm/s; 2.) 1.25 m/s²; 3.) 0.001 J; 4.) 0.001 J

Quiz 12.3 1.) frequency = 0.5 Hz, period = 2 s; 2.) 6.3 s; 3.) 12.4 s; 4.) 1.6 s^{-1}

End-of-chapter questions

1.) frequency = 0.77 Hz, period = 1.3 s; 2.) 0.7 Hz; 3.) 3.5 s; 4.) 70 cm/s, at equilibrium ($x = 0$);
5.) 0.0245 J, $x = -7$ cm or 7 cm; 6.) 11.5 s; 7.) 0.16 Hz

Challenging problems

1.) (a) 0.4 m, (b) 0.8 s^{-1}, (c) 35 cm/s, at equilibrium; 2.) 1.6 s; 3.) (a) 0.05 J, (b) 1 m/s; 4.) (a) 0.2 m, (b) 0.7 s

Chapter 13

Quiz 13.1 1.) amplitude = 2 m, wavelength = 20 m; 2.) $y = (2\text{ m})\cos(0.1\pi x - 0.4\pi t)$; 3.) amplitude = 0.1 m, wavelength = 2.1 m, frequency = 0.96 Hz, wave speed = 2 m/s, direction = left; 4.) -0.1 m

Quiz 13.2 1.) 1200 m/s; 2.) 250, 500, 750 Hz; 3.) 572 Hz; 4.) 286 Hz

Quiz 13.3 1.) 112 Hz; 2.) 3.12, 3.06 m; 3.) 1.6 m; 4.) quiet

End-of-chapter questions

1.) 5×10^{14} Hz; 2.) $y = (2\text{ m})\cos(0.1\pi x + 0.3\pi t)$; 3.) amplitude = 7 cm, wavelength = 0.5 m, frequency = 6 Hz, wave speed = 3 m/s, direction = right; 4.) 286 Hz; 5.) 2, 4, 6 Hz; 6.) 0.1 Hz, 10 s; 7.) loud

Challenging problems

1.) -16 cm; 2.) (a) 4 m, (b) 8 m; 3.) 172, 515 Hz; 4.) small

Chapter 14

Quiz 14.1 1.) 1.63×10^6 N; 2.) 62 m, 3.) 0.2 m; 4.) 0.754 m

Quiz 14.2 1.) 0.05 N, up; 2.) 0.06 m; 3.) 16 m/s; 4.) 0.01 m/s

Quiz 14.3 1.) 114 kPa; 2.) 2.3×10^5 Pa; 3.) up; 4.) 110.5 kPa

End-of-chapter questions

1.) (a) 3 atm, (b) 0.08 m^3; 2.) (a) 1.79×10^5 Pa, (b) 1.88×10^5 Pa; 3.) 2 m/s; 4.) (a) 20 m/s, (b) 2.5×10^5 Pa; 5.) 17 m/s

Challenging problems

1.) 129 N; 2.) 96 mm Hg; 3.) 7 mm; 4.) (a) 1.9 m/s, (b) 0.8 cm

Chapter 15

Quiz 15.1 1.) 2.7 g; 2.) 22.4 L; 3.) 624 J; 4.) pressure = 3 atm, energy = 456 J

Quiz 15.2 1.) 132 J; 2.) 125 J; 3.) 8 J; 4.) 202,600 J

Quiz 15.3 1.) 0.56 (56%); 2.) 0.27 (27%); 3.) 1.5×10^{-5} J/K; 4.) -0.02 J/K

End-of-chapter questions

1.) (a) 0.25 m^3, (b) 37,400 J; 2.) (a) 0.08 m^3, (b) 12,470 J, (c) 20,780 J; 3.) -250 J; 4.) (a) 0.34 (34%), (b) 17 J; 5.) 2.9×10^{-3} J/K

Challenging problems

1.) (a) 10 L, (b) decrease; 2.) 280 K; 3.) 1.0 mole; 4.) 8 J

Made in the USA
San Bernardino, CA
25 August 2015